T0316644

Local Experiences of Mining in Peru

This book uses a multimethod approach to examine local experiences of contemporary mining development in the Peruvian Andes, creating an understanding of the transformations that rural societies experience in this context.

Mining is a major component of economic growth in many resource-endowed countries, whilst also causing mixed social, cultural, and environmental effects. Most current literature on contemporary mining in Peru is largely focused on conflict; however, in this text, the author takes a differing approach by examining the experiences of families in the vicinity of Rio Tinto's La Granja exploration copper project, Northern Peru, an area with great significance due to the mining investment and development, which has taken place over the past 25 years. The book first provides a critical discussion about production of space theories, and debates on spatial mobility, highlighting their relevance to understanding large-scale mining developments, especially in the Peruvian Andes. The following chapters analyze spatial transformations mining development has prompted, focusing on four axes: access to space, production, mobility, and representations of space. A comprehensive narrative is constructed drawing on diverse voices and perspectives, including those of family heads and their partners, local leaders, company employees, and social scientists. The book concludes by discussing how the findings challenge some of the current accounts of the social effects of mining development on rural communities and pose significant implications for sustainable development programs and place-based practices.

By taking an interdisciplinary approach, this book will appeal to a wide audience including geographers, social anthropologists, and social scientists interested in the social effects of mining as well as researchers interested in current Latin American Studies and Rural Development.

Gerardo Castillo Guzmán is an Associate Professor of Anthropology at the Pontificia Universidad Católica del Perú (PUCP), Coordinator of the Anthropology at the City Research Group at PUCP and Honorary Research Fellow at the Centre for Social Responsibility in Mining at the University of Queensland's Sustainable Minerals Institute, Australia.

Routledge Studies of the Extractive Industries and Sustainable Development

Governance in the Extractive Industries
Power, Cultural Politics and Regulation
Edited by Lori Leonard and Siba Grovogui

Social Terrains of Mine Closure in the Philippines
Minerva Chaloping-March

Mining and Sustainable Development
Current Issues
Edited by Sumit K. Lodhia

Africa's Mineral Fortune
The Science and Politics of Mining and Sustainable Development
Edited by Saleem Ali, Kathryn Sturman and Nina Collins

Energy, Resource Extraction and Society
Impacts and Contested Futures
Edited by Anna Szolucha

Regime Stability, Social Insecurity and Bauxite Mining in Guinea
Developments Since the Mid-Twentieth Century
Penda Diallo

Local Experiences of Mining in Peru
Social and Spatial Transformations in the Andes
Gerardo Castillo Guzmán

Local Experiences of Mining in Peru

Social and Spatial Transformations in the Andes

Gerardo Castillo Guzmán

Routledge
Taylor & Francis Group

LONDON AND NEW YORK

from Routledge

First published 2020 by Routledge

2 Park Square, Milton Park, Abingdon, Oxon OX14 4RN

605 Third Avenue, New York, NY 10017

Routledge is an imprint of the Taylor & Francis Group, an informa business

First issud in paperback 2021

Publisher's Note

The publisher has gone to great lengths to ensure the quality of this reprint
but points out that some imperfections in the original copies may be apparent.

British Library Cataloguing-in-Publication Data
A catalogue record for this book is available from the British
Library

Library of Congress Cataloging-in-Publication Data
Names: Castillo, Gerardo, 1971– author.
Title: Local experiences of mining in Peru: social and spatial
transformations in the Andes/Gerardo Castillo Guzmán.
Description: Abingdon, Oxon; New York, NY: Routledge, 2020. |
Series: Studies of the extractive industries | Includes bibliographical
references and index. |
Identifiers: LCCN 2019050541 (print) | LCCN 2019050542 (ebook) |
ISBN 9780367258863 (hbk) | ISBN 9780429292255 (ebk)
Subjects: LCSH: Mineral industries—Economic aspects—Peru. |
Mineral industries—Social aspects—Peru. | Mines and mineral
resources—Peru.
Classification: LCC HD9506.P42 G89 2020 (print) |
LCC HD9506.P42 (ebook) | DDC 338.20985—dc23
LC record available at https://lccn.loc.gov/2019050541
LC ebook record available at https://lccn.loc.gov/2019050542

ISBN: 978-0-367-25886-3 (hbk)
ISBN: 978-1-03-217494-5 (pbk)
DOI: 10.4324/9780429292255

Typeset in Times New Roman
by codeMantra

To Jorge (*in memoriam*) and Rosa, my parents
To Laura and Maga

Contents

Illustrations

Figures

Tables

Acknowledgments

Like the landscape, a book could hide the many people who make it possible. I would like to explicitly recognize their priceless support.

When I finished my doctoral dissertation in 2015 and returned to Peru, my home country, I thought that milestone of my academic life was completed. David Brereton and Daniel Franks, my advisors, proved I was wrong. Daniel stubbornly encouraged me to convert the dissertation into a book and enthusiastically supported its publication. David has been an inspiring and generous presence along these years. His conversations and guidance have provided me with a broader, deeper, and nuanced comprehension of the social processes mining development contributes to shape. He has carefully gone through many cumbersome drafts and has offered me fresh and sharp ideas. David and Daniel have enriched my life, both academically and personally.

Not only were the members of the families in the study, along with the men and women in La Granja, extremely generous in sharing their time and experiences, but they also offered me their friendship. I thank every single one of them.

Geographer Marilyn Ishikawa elaborated the graphs that are included in this book. I extend my most sincere appreciation to her.

Glenda Escajadillo edited the manuscript with care, dedication, and professionalism.

A research grant from the Pontificia Universidad Católica del Perú (PUCP), my host institution, allowed me to dedicate extra hours to writing duties. In addition, funds from the PUCP's Vice-Chancellor of Research—FONCAI were allocated to elaborate the maps and the edition of the manuscript.

Laura Soria has been instrumental in the writing of this book. I shared many ideas with her; her experience in gender-related topics and her understanding of contemporary Peruvian society have been extremely valuable. Her permanent and everyday presence, along with Margaret Mead II (Maga), has been my greatest support. To Laura, I give all of my love and gratitude.

1 Introduction

New light for old ghosts

1.1 A story

It was in the early hours of July 1997 that Alfredo Granda remembers having left his home in northern Peru with his wife, his elderly parents, and his six children, the youngest barely three weeks old.[1] They did not leave with hope but with resignation. When Alfredo left the village, there were only seven houses still standing, and the sixty other houses had already been knocked down by Cambior, a Canadian mining company that had bought the lands from local families for the purpose of exploiting one of the largest copper fields in the world. As many others had, Alfredo and his wife, Mercedes, resisted selling their land and house for several months. However, pressure from the company and the government had forced them to sell. Escorted by the security staff and a lawyer, Mr. Melgar, an engineer and the company's strongman, had threatened them. If they did not sell and leave, the army would evict them with no compensation whatsoever, and, in addition, they would be accused of being terrorists. The state had already closed the village first-aid post and school. Over 250 families sold their land.

In a small secondhand truck bought with money from the sale of his land, Alfredo and his family left behind the green inter-Andean valley in the Cajamarca region. Two of his brothers had tried to convince him to move to the jungle, where they could acquire land by cutting down the forest trees and growing coffee. Drug trafficking and lack of safety discouraged him. He would rather have moved to the coast, near the cities, where his children could study and find jobs. Two months earlier he had gone to Ojo de Toro, a rural area three hours away from Chiclayo, the main city, and the regional trade center where hundreds of migrant families from Cajamarca have settled. There he bought a plot of land and left instructions and money for the construction of a house. They then departed for that semiarid locality.

The reception they received on arrival was hostile, and nothing turned out the way they thought it would. They were scammed and the house they had paid to be constructed did not exist; they had to move temporarily into a straw mat shack. In hot weather, the harvest cycle is different. In addition, Alfredo had never grown crops with the use of irrigation. His first corn harvest failed. The little money they still had from the compensation was soon gone, and with neither family nor friends, the Grandas had nothing to eat. Alfredo worked for food on his neighbors' small farms. The following year, the heavy rains that were caused by the El Niño phenomenon finally ruined the family. The fields flooded and the area was declared an emergency zone. The family was provided with shelter by a religious organization, and for several months they lived in tents that were set up in a public school. The rainfall, the floods, and the lack of potable water brought about malaria and cholera epidemics. Alfredo's father and the youngest child died. In despair, the family decided to go back to La Granja. As they had no land, they felled and squatted in a portion of the nearby forest, where they built an adobe house and grew cassava, potatoes, and corn.

The mining project that had been responsible for their displacement did not prosper, and Cambior sold its rights and land to BHP Billiton, a multinational company that also eventually chose not to proceed with the project. BHP Billiton returned the mining concession to the state and offered the land for sale back to the original owners at a reduced price. With the help of a brother and a son-in-law, Alfredo was able to buy a portion of his old land. Little by little, other families returned and started to rebuild their old lives.

In 2006, yet another mining company, Rio Tinto, received the concession rights from the state and started new exploration works. The company established a program that employed the local population by the dozens, and families started many small businesses to provide services to the company and its workers. Five of Alfredo's seven siblings also returned. His eldest daughter, Margarita, opened a laundry. One of his sons went into business with a cousin to start a small hotel and a snooker room, where young people from La Granja gather in the evenings to play. They no longer drink cane liquor, like their fathers used to; now they drink beer and listen to reggaeton music rather than the Andean music of the old-fashioned *huaynitos*. As the village now has electric power, the Grandas gathered at home every evening to watch soap operas and game shows by satellite signal, and they kept in touch with each other by means of mobile phone. Along with many other people, Alfredo thought that the company would soon start buying land for the project. That is why he bought some land from his neighbors and divided it among his children and their families. Each child built their own house and replaced donkeys and horses with motorcycles and pickup trucks. With mixed feelings of nostalgia and optimism, Alfredo and Mercedes witnessed financial effervescence being experienced by the village like no other time in its history. They also felt fear and uncertainty, however, about what the future might bring.

1.2 Mining and social change in rural communities: a disruptive force

Large-scale mining is a major force in the transformation of rural societies around the world (Bebbington & Bebbington 2018). On the positive side it significantly contributes to the economic growth of many resource-endowed countries, by providing significant foreign investment and taxation to the state at the national level (UNCTAD 2007; Slack 2009). However, well-documented evidence indicates that mineral development often also causes and contributes to mixed social, cultural, environmental, and economic outcomes for local communities (McMahon & Remy 2001; Rosser 2006; Damonte 2008; Bainton 2010; Bebbington & Bury 2013b; Brain 2017).

Complex outcomes for rural regions impacted by mining development have included the creation of new and relatively well-paid jobs outside farming activities, the provision of infrastructure, substantial inflows of capital, local inflation, in-migration and out-migration, imbalances in the women-to-men ratio, changes in land use and ownership, political struggle and fragmentation, and cultural clashes. As some researchers examining development projects and modernization processes have shown (Ferguson 1990; Escobar 1995; Coronil 1997), these transformations often lead to uneven and contradictory forms of social and economic development, with gender relations, in particular, being one of the areas in which inequality is heightened (World Bank 2001; Castillo & Soria 2011). At the same time as researchers have been documenting these impacts, a burgeoning literature has developed from the corporate social responsibility perspective seeking to address the sustainable mining "oxymoron" (Franks 2015).

Research analyzing the perspective of the local populations and their experiences and responses to mining activities has been important since the second half of the 20th century (Godoy 1995). Current research mainly focuses on the resilience of communities and the strategies employed to resist and confront mineral developments.[2]

Often using the methodological and theoretical framework of political ecology, these studies have addressed imbalances of power and examined the multiple political and symbolic resources that local populations use to oppose mining (Castillo 2006; Bebbington 2007; Damonte 2007; De Echave et al. 2009). Authors from this tradition have stressed the threat that mineral development poses to traditional agrarian systems (Bury 2004; Salas 2008; Gil 2009; Torres 2013). However, the learnings and practices that local populations deploy in order to deal with mining transformations, which do not necessarily imply resistance and social conflict, have been less explored. A narrow understanding of local responses severely limits our comprehension of the social transformations that may occur because of mining development in rural societies.

1.3 Study focus

This book utilizes a case study of a community in Peru to challenge and extend the existing literature on the impacts of mining on rural communities. It seeks to understand the spatial transformations that rural populations experience and actively shape in response to large-scale mining development in the Peruvian Andes. The focus is on the experiences and responses of the families near La Granja copper project in Cajamarca, in the northern Andes of Peru, since mining activities began, around 25 years ago. In particular, the book centers on the examination of local experiences related to land and housing (*access*), farming and nonfarming activities (*production*), and migratory history (*mobility*), as well as the images of the past, present, and future La Granja (*representations*) that women and men living in the vicinity project have constructed since 1994.

I argue that through diverse experiences and practices, local populations are helping to significantly transform the society in which they live. The underlying proposition is that spatial transformations prompted by current mining development in Peru exhibit features that depart from conventional accounts of social change in rural global North. First, there are signs of higher market integration at regional and national scales between rural and urban areas, although in contrast to the classic Western experience, farming productivity gains do not accompany this process. The increase of paid labor in nonfarming activities and the consumption of external goods are the factors producing this market integration. The second feature is that the resulting urbanization process is not binary or closed from rural to urban lives, but a mixed and fluid one in which families use their networks in order to bridge both spaces. Third, social relations have not necessarily become more individualistic or anonymous; kinship and social networks remain central to individual lives, although some social relations and identities are increasingly challenged, especially gender ones. Finally, gender and age, as well as collective and individual experiences and interests, strongly shape the construction of social representations of the city and the countryside.

These larger outcomes can be understood as a social transformation process from the bottom, where kinship and local networks act as a distribution system and safety net. In other words, an important part of the distribution of goods and services and many of the decisions regarding mining development at a local level are regulated by extended kinship networks and not exclusively by state institutions or company policies. This perspective is consistent with, and to some extent complementary to, current approaches that emphasize the fragility of the state and the hybridity of the political order (Boege et al. 2008). In a fragile state, the institutions are weak and have serious difficulties providing services, goods, and law and order to populations beyond the central capital. In addition, and partly because of the absence of the state in a great part of the territory, citizen participation and representation are inadequate, generating a lack of state legitimacy from the people's perspective (Boege & Franks 2012). This hybrid order is characterized by its combination of traditional social structures, elements from the Western model of the state, and organizations and movements originated by and reacting to globalizing forces.

1.4 Theoretical lens

The book uses the theoretical framework of "production of space", developed by the French geographer Henri Lefebvre (1991) as a tool for the understanding of social transformations propelled in the context of natural resource development. This theoretical lens has the advantage of directly addressing the issue of agency in the production of social realms, incorporating both temporal and spatial analyses for the understanding of social phenomena, linking diverse spatial scales, and encompassing symbolic, power, social, and economic elements within a single framework.

The concept of "production of space" stresses that space is not merely a natural realm where human activities occur, but that these activities produce it in a dialectic process. In this sense, each society or, more precisely, each way of organizing production in human history (the "mode of production") creates its own space. Therefore, space is deeply embedded in the social realm and encompasses material as well as social and ideological elements. Space is a social production and, therefore, historical processes shape it.

Lefebvre states that space is not only composed of physical attributes. Space is a social product that different actors experience, perceive, and imagine in conflicting and contested ways. In the capitalist mode of production, it implies three intertwined dimensions: spatial practices, representations of space, and representational spaces. The first refers to the circulation of people, goods, services, and capital, the social networks and daily routines, and the creation of spatial zones (i.e., private or community property). The second relates to the vision that the producers of space have of it (maps, spatial hierarchies, borders, or forbidden spaces). The final dimension describes the way that the users of space and artists imagine and propose alternative visions and symbols for its use (through pop culture and media, popular spectacles and graffiti, or creation of symbolic and contra-hegemonic capital). In addition, these spatial dimensions evolve with capitalist expansion; that is, the expansion of capital sets in motion different arrangements regarding how space is produced, experienced, ordered, measured, conceived, imagined, and contested.

At the same time, I consider there are some limitations in Lefebvre's framework for the understanding of spatial transformations in industrial capitalist societies. Lefebvre elaborated his ideas within the context of the second half of 20th-century French urban development; that is, in a highly industrialized, state-centered, and regulated setting where the state and urban development firms are the producers of space par excellence, exercising great power over citizens or users of space. By contrast, in societies characterized by weak state institutions and regulations, and incipient capitalist development, lay people also become key producers of space. This is the case, for instance, of the urban growth of Lima resulting from the influx of millions of migrants from the 1940s to the 1990s (Calderón 2005). Therefore, I avoid the sharp distinction between producers and users of space proposed by Lefebvre. Instead, following the insights provided by political ecology, I distinguish between state agencies, the private sector, and local populations, organized or otherwise and crossed by different categories of class, gender, age, or ethnicity. Where prevailing accounts of social transformations in Peruvian Andes overlook the perspective of the local populations, this book explicitly seeks to provide a voice to men and women for the explanation of social change in the context of mining development.

1.5 Study location

The research area is situated in the Cajamarca region of Peru. Cajamarca hosts many large-scale mining projects and in the past two decades has received significant investment for mineral development. This study particularly focuses on the La Granja project,[3] one of the largest undeveloped copper projects in the world and owned by Rio Tinto. The project site, which has a lease area of 7,400 hectares, is situated 2,000 m.a.s.l on the western side of the subtropical Andes in northern Peru, in the district of Querocoto, province of Chota (see Figure 1.1).

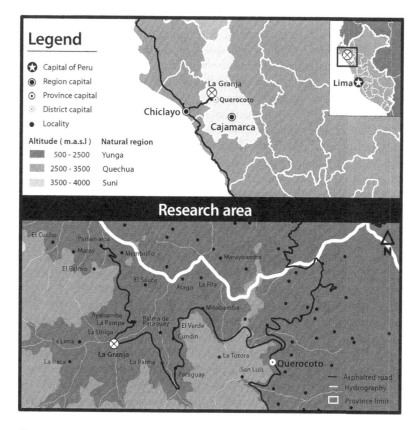

Figure 1.1 Map of the case.
Source: Castillo (2015, p. 44).

The main localities or *caseríos* in the direct area of influence of the project are La Granja, La Iraca, La Pampa, El Sauce, Paraguay, Cundín, La Fila, El Verde, and La Palma. All of them belong to the district of Querocoto.

The *caserío* of La Granja is the nearest locality to the mining site and is which takes its name from. La Granja is nearly 1,000 km from Lima. Although La Granja is in the region of Cajamarca, its main connection is not with the city of Cajamarca but with Chiclayo, a major commercial city on the northern coast of the country and located in the region of Lambayeque. After infrastructure improvements to the road surface for the mining project a bus journey from Chiclayo to La Granja could take around 15 hours.

1.6 Methodology and data sources

The book is based on my doctoral dissertation (Castillo 2015), which explored the dimension of access to space, production in the space, spatial mobility, and spatial representations through different voices and perspectives: family heads and their partners with different spatial experiences, local leaders and qualified informants, company employees, and social scientists. The study also made use of a diversity of data sources: in-depth interviews, ethnographic observations, socioeconomic secondary data, and local fiction. The fieldwork, which comprised three stages totaling 85 days of participatory observation, was conducted in 2013 as a "multisite ethnography" (Marcus 1995) in order to deal with the geographical mobility that mining development implies. The fieldwork involved interview sessions and ethnographic observations in four locations: the La Granja area; Querocoto, the district capital; Chiclayo, the main coastal city in the region; and Ojo de Toro, a rural coastal locality in Lambayeque Region, as Figure 1.2 shows. It included in-depth interviews with members from 14 families[4] and semistructured interviews with 24 qualified informants in La Granja, Chiclayo, and Lima, the country's capital.[5]

At the time I conducted the fieldwork, there was a very active exploration program in the area. However, from 2015, Rio Tinto made the decision to put the project on hold and the activity has significantly slowed down. As of 2019, it is unclear whether and when Rio Tinto will move into the development phase, or if the company will seek to withdraw. The company's decision will depend on China's copper demand, but there are also technical characteristics of the project that make it difficult to develop (e.g., high levels of arsenic to contend with). Meanwhile, the state has extended the exploration rights to Rio Tinto for seven years, from 2017 to 2024 (Fernandez 2019).

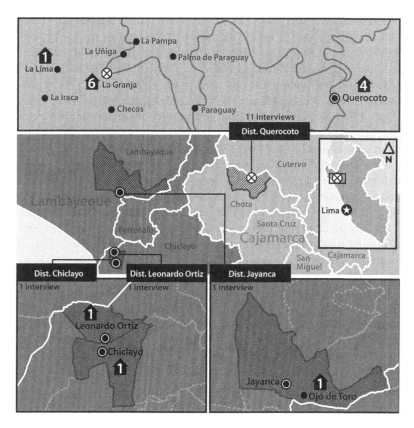

Figure 1.2 Location of interviewed families.
Source: Castillo (2015, p. 31).

The analysis presented here places a strong emphasis on the experiences of adult family members, both male family heads[6] and their partners, establishing an interplay between collective and individual actors, between family and individual experiences and the link to broader social-economic conditions and the institutional settings that have evolved over time (Offen 2004).

This focus on nuclear families[7] sets this study apart from the work of other researchers who focus on the peasant or *campesino* community as being the center of analysis. Anthropological research in the Andes (Mayer 2002) has made clear the importance of extended family networks for individuals and has shown the usefulness of a family analysis. Not only are individual behaviors and actions strongly shaped by kinship relations, but also the production and circulation of goods and services takes place in the space of the household, which mediate other levels of collective action. Although for many Andean populations the *campesino* community is the formal space of collective decision-making and the legal owner of the land, in practice, the nuclear family has become the central unit of production and the recognized holder of the land (Kervyn & Equipo del CEDEP AYLLU 1989; Mayer 1996). While the *campesino* community possesses the legal titles as a collective owner, private property exists, and each family appropriates the benefits of the farming activities of their own land.

This is especially the case for rural society in Cajamarca. Due to historical reasons, rural populations in the region have not organized themselves through the *campesino* community institution (Deere 1990; Taylor 1994). Instead, they have mainly organized around relatively independent families grouped in villages (*caseríos*) and are part of community protection organizations (*rondas campesinas*). These community protection organizations first emerged in the 1970s to fight against cattle rustling and evolved in the 1990s to successfully resist the incursions of the Maoist group Shining Path (Taylor 1997). Nowadays, the *rondas campesinas* are significant actors in the regional political scenario and are a powerful mediator between families and external actors, including mining companies and the state. However, by no means does the *ronda campesina* replace families in making the most important decisions regarding livelihoods (for instance, farming management, migration, or investments).

Although nuclear families are the primary research units, it is misleading to consider them as homogenous entities. They not only greatly differ among one another, but also internally. To address this, I have developed a typology of five groups of families according to the following spatial criteria: (i) those who refused to sell their property to Cambior and did not resettle (resistant); (ii) the families who sold their land but returned to La Granja when BHP Billiton offered the land back (returnee); (iii) the families who migrated to the village when new economic opportunities emerged with the arrival of Rio Tinto (opportunistic); (iv) the families who did not return to La Granja form the fourth group (migrant); and (v) the families located in the capital of the district, outside the direct area of influence of the project (regional).

1.7 Structure of the book

After this introduction, Chapter 2 provides a historical outlook of spatial transformations and mining development in the last six centuries in Peruvian Andes. Chapter 3 narrates the history of La Granja as an agrarian space and describes its continuous fragmentation over the last one hundred years. Chapter 4 examines how the *hacendado*, the state, mining companies, and local families have used political, market, and kinship mechanisms for accessing land and urban properties. Chapter 5 interprets the shift and complementarity between farming and nonfarming activities among *Granjino* families. Chapter 6 explores the migratory experiences of local families and the complex spatial networks they have built. Chapter 7 analyzes how age, gender, and family history interact for the construction of past, present, and future representations of La Granja's social space. The final Chapter 8 reflects upon the specificities of the examined case to challenge some of the current accounts of the social effects of mining development on rural communities—regarding changes over land access and land use, the fluidity of urban–rural mobility, the functioning of strong social networks in the context of weak institutional settings, and the configuration of hybrid identities—and outlines a research agenda.

Notes

1 The names of these characters and the brief story of their saga are fictitious. They portray, however, some of the dramas experienced by many families in La Granja.
2 See the specialized reviews written by Ballard and Banks (2003), Bridge (2004), Damonte and Castillo (2011), and Gustafson and Guzmán Solano (2018).
3 For a description of the project development's history, see Chapter 3.
4 Three resistant families, two returnee, two opportunistic, four migrant (one in urban Andes, two on the urban coast and one on the rural coast locations), and three regional, which totaled a number of 280 hours of conversation. For a brief profile of each interviewed family, see Table A.1 in the Appendix.
5 Fifteen local informants, five academics and social consultants, and four employees of the Rio Tinto La Granja (RTLG) project.
6 I use the term "male family head" for two reasons. First, the continuity of a patriarchal system permeates the fabric of the society in the research region (Deere 1990) as much as other rural Andean areas (De la Cadena 1991; Hamilton 1998). Though increasingly challenged, a patriarchal system is still present on issues surrounding access to the land, productive and reproductive labor, social and geographical mobility, representations of space and society, and so on. Thus, the term attempts to reflect an inner perspective of intra-household power relations. Second, and even more importantly, when I asked who the appropriate person was to talk about the different issues of the research, I was generally directed to an adult male. Conscious of my position as adult, male, and outsider, I did not presume to question these gender relations. Only after becoming acquainted

did I begin to converse with women. My prolonged stay in the research area, living with a well-known local family, facilitated the process. The conversations were conducted in both public spaces and their houses, usually without the company of their partners, which allowed the female informants more freedom to share their views and experiences.

7 Most economic surveys, especially in developing countries, define a household as "a group of people who live together, pool their money, and eat at least one meal together per day" (United Nations cited in World Bank 2001, p. 150). However, due to conditions of fluid migration, I prefer the use of "nuclear family" as unit of our analysis. For the purposes of this research, I define nuclear family as a group of people who recognize themselves as bonded through kinship—two generations before and two after the male family head—and do not necessarily live in the same place. Thus, for instance, a girl living in Chiclayo, and her parents, who live in La Granja and send her money for her studies, are part of the same nuclear family. In brief, in a single house, different nuclear families could cohabit and, in addition, a single nuclear family may occupy different houses. Adopted children are included within the concept of nuclear family.

2 Spatial transformations and mining development in Peruvian Andes

The Andes are one of the largest mountainous ecosystems in the world and goes from Patagonia (Argentina and Chile) in the south to the Caribbean Sea (Colombia) in the north. In this book, I focus on the Central Andes, the area that comprises the current borders of Bolivia, Peru, and Ecuador, an area that has been the cradle of major civilizations and settlements. This mountainous ecosystem has shaped, and in turn, been transformed and managed by human activities (Dollfus 1981); transformations in the spatial production of the Central Andes that can be summarized in six major periods over the last 600 years.

The Inca State conducted the first spatial transformation between the late 15th and early 16th centuries. In the Andean region, altitude is the most significant variable that explains variation in farming production and access to natural resources. Therefore, within a range of relatively few kilometers, populations could access seafood, beans, corn, potatoes, and other Andean tubers, and pastures for camelids, as the altitude increases from the sea level to the high peaks. In addition, the eastern slopes of the mountains lead into the Amazon basin where fruits, coca, and other tropical products and animals are found. In a pioneering and influential work, John Murra (1975)[1] postulates that within that ecological setting, which provides both severe constraints and opportunities, the best strategy that Pre-Hispanic populations deployed in order to maximize their access to natural resources and micro-climates was the so-called "Andean ideal of vertical control of a maximum number of ecological zones". The implementation of this ideal means that using kinship bonds, ethnic groups allocate different household units at different ecological niches to access different products. This ethnic archipelago, as shown in Figure 2.1, did not imply the exclusive control of the territory but the coexistence of diverse groups in a shared space.

Vertical control of ecological niches

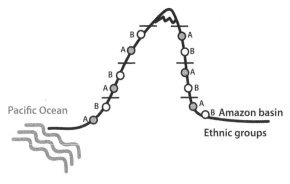

Figure 2.1 Pre-Hispanic political and spatial control of the Andes.
Source: Castillo (2015, p. 51).

In the same work, Murra explains that the Inca State used, adapted, and enhanced previous kinship strategies of spatial production and control for its own purposes of expansion. Therefore, the Inca State gained direct control over large farming lands and obligated the population to work in them. In addition, it implemented a policy of forcing specific groups[2] to abandon their places of origin to be allocated in remote areas with the purpose of fulfilling production and political needs. The population mostly cultivated potatoes, the main staple and core of the daily diet, and the Inca's bureaucracy strictly controlled the farming of maize[3] as an exclusive crop within the state's lands. Hence, the Inca State functioned as an enormous octopus that controlled and centralized farming products to then distribute them for public (e.g., maintenance of a large army and bureaucracy) and caste (*panacas*) purposes in a political system without market mechanisms.

The second transformation of Andean space took place during the 17th century. Originating in the colonial experience, it signified a dramatic change for the native Andean people and has had enduring consequences that still shape social and economic conditions today.

Prior to the Spanish conquest, the number of people inhabiting contemporary Peru has been estimated at between 9 and 15 million (Cook 2010). However, following the conquest years of war, famine, social disintegration, and, more importantly, the spread of new diseases (namely flu, measles, and smallpox) the population shrank to just one million (Wachtel 1976; Cook 2010).

The Spanish colonization of America (over the Andean and Aztec societies) radically differs from other experiences (i.e., the British colonization in North America) in at least two key aspects. First, the newcomers were not settlers specifically looking to farm their own land. Embedded within the social relations of feudal Spain, they performed as conquerors seeking to obtain noble titles over the land and secure servants to work it for them. Second, the ecological conditions the Spanish found were very different from the peninsular ones, so instead of merely importing techniques and species from Europe, they had to adopt native production systems. Thus, for the Spanish colonial enterprise, the control of scattered indigenous labor was crucial for redirecting it to the functioning of the farming, textile, and mining industries. Hence, under the rule of Viceroy Toledo,[4] the colonial system created a division between the Republic of Indians and the Republic of Spaniards. This division implied a reorganization of the population, the territory, and the social relations in two different though connected entities: the indigenous community (*comunidad de indios* or *reducciones*) and the large-state property (*corregimiento*) (Spalding 1974). The inhabitants of the newly created *comunidad de indios* were evangelized into Catholicism, schooled in Quechua, and dressed according to rural Castilian customs; thus, the cultural landscape became homogenized. Marginal and scarce lands were collectively secured for the *comunidad de indios* with the purpose of enabling the physical reproduction of the indigenous people (as a labor force assigned to specific Spanish *corregimientos* or mines) and the payment of taxes to the Crown (Fuenzalida 1970a).

The third milestone signified the end of the *corregimiento* system in the aftermath of the Túpac Amaru rebellion in the late 18th century. The uprising led by indigenous noble and entrepreneur, Túpac Amaru, was the most important indigenous revolt in the Peruvian Viceroyalty and was prompted by the attempts to implement the Bourbon reforms in the Spanish colonies (Fisher 1971). Two major consequences of the rebellion were the limitation of the most oppressive features of the *corregimiento* system (and with that, some weakening of the regional landlords) and the disappearance of the indigenous leaders (*caciques*) and nobility (O'Phelan 2012).

The fourth period was one of reappropriation and expansion of the large-state property system (*hacienda*). This period began after the Independence Wars, around the 1840s, and lasted until the 1950s. This process of property expansion was driven by capitalist production and signaled the dominance of English and American interests in the country. Thus, for instance, in the Southern Andes a productive and commercial circuit connecting peasants, *criollo*[5] landowners, and British traders was established for the provision of wool to textile factories in England (Flores Galindo 1977).

Of course, this capitalist expansion at the cost of communal lands was not without resistance. Peasant resistance was relatively successful where the state and regional elites were weak (Mallon 1983) and by the late 1950s, it became evident that there were limits to that kind of modernization. During the 1960s, the *campesino* movement became acquainted with leftist ideologies and leaders, from which discursive and organizational tactics were borrowed. A major *campesino* revolt and appropriation of lands in the coffee plantations of Quillabamba, on the eastern slopes of Cusco (Hobsbawn 1974), demonstrated the need for land redistribution.

Consequently, the fifth transformation was conducted by the military leftist government of Juan Velasco and began in 1969. Though short in duration, it led to one of the most radical agrarian reforms in Latin America and implied the destruction of the *hacienda* system, both on the rich export-oriented cotton and sugar plantations of the coast and the semifeudal *haciendas* of the highlands (Matos Mar & Mejía 1984; Eguren 2006). Under the slogan of *"campesino*, the landlord will not feed from your poverty any longer", this reform not only destroyed the prevailing land property system of the time, but also attempted to create communal forms of property—Agrarian Production Co-Ops in the coastal areas and Social Interest Agrarian Societies in the highlands. The latter was heavily resisted by the *campesinos* who sought to control the land by their own and not through the state bureaucracy.

Thus, backed by an uncommon alliance of leftist political parties (Revolutionary Vanguard and the Revolutionary Communist Party), the social democrats (Popular Action), and the liberals, the *campesinos* took the land by the force and divided it up into individual plots for each family (Rénique 2004). This was the sixth major transformation of the Andean space and marked the end of decades of struggle over the land for agrarian purposes.

The 1980s were marked by dramatic economic and political phenomena for the country. Significantly, Peruvians suffered the insurgence of Shining Path, a radical Maoist group that prompted a war with the state, social movements, and civilians at the cost of close to 70,000 deaths (Comisión de la Verdad y la Reconciliación 2004) and on the highest inflation records in world history, which devastated the economy. Despite the magnitude of these phenomena, land use and land ownership remained relatively stable for around two decades.

More recently, the scenario has begun to change after the liberalizing policies were implemented in the early 1990s. These policies have been instrumental in the emergence of a large mining boom in the country that has lasted more than 20 years and are producing a cycle of major transformations in Andean societies.

From being a marginal activity during Pre-Hispanic times, mining has been of central importance in Peruvian society over the last five centuries. However, this centrality has completely shifted from being an agent of economic integration to a major source of social conflict. Certainly, mining has been crucial to the Peruvian economy and society since colonial times. As Assadourian (1982) has shown, the exploitation and the commercialization of silver from the large deposits of Potosí organized and integrated a regional market, which now includes Ecuador, Peru, Bolivia, Chile, and northern Argentina. The decline of Potosí's production and the Independence Wars of the late 18th century seriously affected mineral production. However, as Contreras (1995) has argued, mining continued to be an important agent in regional economies with the opening of new mine sites in the territory. For instance, some authors (Manrique 1987; Deustua 1994) have explained how mining activities have provided economic surplus for the recovery of cattle and farming activities following the Pacific War in the 1890s.

Nevertheless, the advent of the 20th century was a turning point, which heralded the emergence of modern mining. As Brundenius (1972) and Manrique (1987) describe, modern mining implied three major changes: the shift from a labor-intensive to a capital-intensive industry; the entrance of foreign interests into the sector, and the decline of Peruvian capitals in large-scale mining projects; and the establishment of the enclave scheme. These combined factors produced a new system in which mineral development was directly linked to international markets without the necessity for local/regional products or workers. In other words, mining ceased to be an important agent in the integration of regional goods and labor, and instead became a disruptive actor among local populations. For the case of mining in the Central Peruvian Andes, Helfgott's detailed work (2013) illuminates the profound and devastating consequences, for the local rural communities, of the major shift in the labor, production, and housing regime of mining operations. Workers, traditionally recruited from the surrounding rural villages in a seasonal cycle that created what was called the "peasant-miner" (Long & Roberts 1984; DeWind 1987), became redundant as a capital-intensive regime produced a labor surplus from a previous labor shortage.

Since the 1990s, the mining boom has contributed greatly to the sustained growth of the Peruvian economy (UNCTAD, WB & ICMM 2008). For instance, during the peak of the mining boom, minerals account for around 60% of the country's total exports (Castilla 2012). This mining expansion has been supported by international financial institutions, namely, the World Bank and its private sector arm, the International Financial Corporation. In what was regarded as a new era for the mining sector, these institutions promoted the design and implementation of the so-called "New Extractive Industry Strategy" (Arellano-Yanguas 2011). This strategy aimed to foster social development through two measures: first, the distribution of significant state tax revenues into mining regions; and, second, the promotion of more active participation of mining firms in social development initiatives following the practices of corporate social responsibility. The results, however, have not met the state's goals and, currently, mining is a major source of social conflict in the country.

Much of the social conflict is triggered by the negative socioeconomic consequences that mining operations create at the local level. Perhaps the most notorious impact is the—factual or perceived— environmental degradation, specifically the effects on the quantity of available water and its quality. The need of large-scale mining operations for water directly collides with local farmers' interests and fears. In the context of poor environmental standards and weak state surveillance mechanisms and institutions, water is one of the most contested issues of mining operations (Bebbington & Williams 2008; Himley 2014). In addition, land is also significantly affected; not only the quality of the soil in areas directly adjacent to the mine sites, but more importantly, because of the changes in use and ownership in larger areas that are fostered by the development of the mineral operations (Bury 2004). In some cases, changes in land use and land ownership have resulted in processes of migration or involuntary displacement (Castillo & Brereton 2018b; Szablowski 2002).

Additionally, the creation of mining-related jobs and the injection of relatively significant amounts of money into local economies have generated two unforeseen effects: inflation and the shortage of workers for farming activities (Viale & Monge 2012). These effects especially hurt vulnerable families who are not engaged in mining activities. The combination of these and other factors has fostered deep transformations in the livelihoods of many rural families, particularly those living in mining regions (Bury & Kolff 2002; Brain 2017) and has, in turn, led to distress and social tension.

The mining-induced production of space in the Andes differs from previous transformations in at least one central issue: it does not necessarily involve the control of farming activities or the use/control of local labor. The focus of this spatial configuration is the control of land and water for mineral resource extraction.

For many rural areas of the Peruvian Andes, mining development is a major driving force of social change, which has not been experienced since the dramatic events of the agrarian land reform of the early 1970s. Viewed in retrospect, Andean societies have undergone a series of major transformations resulting from a way in which top-down forces have organized territory, natural resources, and people. The transformation propelled by large-scale mining development is the most recent in this series. In what follows, we will examine some of these transformations from the perspective and experience of the local families around La Granja project in the northern Peruvian Andes.

Notes

1 Murra introduced diverse variations along the time and perhaps its more conclusive version is contained in the compilation of his work published by the Instituto de Estudios Peruanos and the Pontificia Universidad Católica del Perú (2002). Murra admits variations from the archetypical model, including the thesis of María Rostworowski, which suggest the existence of a horizontal model of territorial control in the coastal areas. For further discussion, review, and expansion of Murra's model, see Masuda, Shimada, and Morris (1985). Of course, these strategies are not only utilized in the Peruvian Andes; see, for example, the work of Geneviève Cortes (2004) who explores migratory strategies of contemporary farming communities from two different ecological niches in Bolivia.

2 This is the case of the *mitmakuna* or "scattered people".

3 Maize was the most prestigious crop since it is the base for the *chicha*, a widespread Andean beer central in any ceremonial activity.

4 From 1569 to 1581.

5 *Criollo* is the name given to Spanish descendants born in the American colonies. Because of their higher levels of education, social ranking, and wealth—as well as freedom from legal restrictions—the *criollos* were able to maintain the highest power and economic positions after independence from the metropolis.

3 The historical production of La Granja

Diverse social, political, and economic factors have produced La Granja as a social space over time. Within this evolving process, mining-driven changes are the last in a series of transformations in this rural Andean space. These transformations are better understood in relation to complex connections with diverse localities, which are part of a major region that connects the coastal city of Chiclayo with its hinterland in the Andes and the Amazon basin. In a fluid process of incoming and outgoing migration, La Granja is a social space composed of family networks that extend beyond the limits of the locality.

La Granja began as a large *hacienda* in the early part of the 20th century, although the lack of official historical records and the limits of local collective memory make it difficult to establish with accuracy its precise origin. The area was previously known as Paltic, the name of the river that waters those lands and forms part of the Amazon Basin, which flows into the Atlantic Ocean. Around 1920, Cecilio Montoya, the first *hacendado*, changed the area's name to that which is used today, La Granja (The Farm). The extension of the original *hacienda*, approximately 200 km², was significantly bigger than the current territory.

The La Granja *hacienda* system made use of pre-capitalist labor and social relations. Thus, the *hacendado* kept the best farming lands and pastures for himself and rented plots to landless *campesinos* from the region who, with their families, migrated to establish lives in the *hacienda*. These *campesinos* (known as *arrendatarios*) and their families become personally bonded to the *hacendado*. In La Granja's case, the immigrants came from populous Cajarmarca provinces, such as Hualgayoc, Chota, and Cutervo (see Figure 3.1). In exchange for land, the *hacendado* received payment both in cash and farming products from the *campesinos*. In addition, the *hacendado* requested personal services of men and women for farming (i.e., cultivating the land or feed

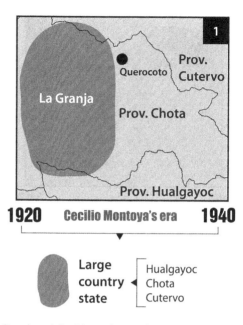

Figure 3.1 La Granja original large *hacienda*.
Source: Castillo (2015, p. 66).

cattle) or domestic activities (i.e., housekeeping of the manor house or weaving of textiles for commercialization in regional markets). These services were not only unpaid and represented a great burden of time, but they were also based on an arbitrary system of immediate demand. In extreme but not unusual cases, the *hacendado* forced the *arrendatarios* to hand over their young daughters for sexual favors under the threat of expulsion from the *hacienda*. Moreover, the *arrendatarios* were not allowed to construct their houses with brick and cement but only of mud and straw, so these constructions could be easily removed or burned down. The *hacendado* provided basic primary education to the *arrendatario*'s children but further education was forbidden. In this labor and social system, the *hacendado* used *mayordomos* (foremen) as intermediaries. Thus, the *mayordomos* oversaw the daily work in the *hacienda*, benefiting from a space of relative freedom but also exercising and reproducing the oppressive conditions for the *arrendatarios*.

Of course, this was an oppressive and servile system. However, in a context of insecurity, extreme rural poverty,[1] and absence of the central state, the *hacienda* system provided protection from banditry and

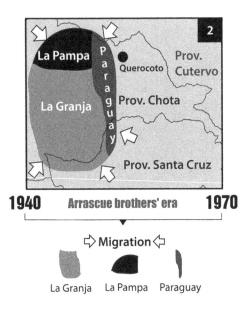

Figure 3.2 First fragmentation of the La Granja space.
Source: Castillo (2015, p. 67).

social violence and a patch of land in which to work and make a living. Thus, the *hacienda* became the space for a paternalistic and patriarchal system of economic exploitation and social protection.

Around 1940, Cecilio Montoya died; and, after a decade, his daughters sold the *hacienda* to the Arrascue, three brothers from the province of Chota: Alejandro, Gilberto, and Wenceslao. The original estate was divided into three parts and Alejandro took possession of La Pampa, Gilberto of Paraguay, and Wenceslao of La Granja (see Figure 3.2).

This period in La Granja's evolution lasted for more than two decades and because of pervasive poverty in the region, there were similar flows of immigration as in the previous age. However, political, social, and economic processes produced dramatic changes that signified the collapse of the *hacienda* system. These processes led to the agrarian reform initiated by the leftist military government of General Juan Velasco Alvarado in 1969, which radically transformed the rural landscape of Peruvian society. Around the end of the 1960s, the signs of social change were evident and, like many other *hacendados* in the country, the Arrascue brothers began to sell part of their marginal

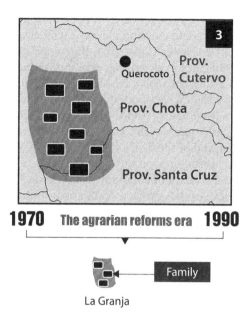

Figure 3.3 The end of the *hacienda* system.
Source: Castillo (2015, p. 69).

lands to their *arrendatarios*. This measure prompted a struggle over access to and ownership of the land and two sides were formed: the *amarillos* (the yellow side) and the *rayados* (the striped side). The *amarillos* were *arrendatarios* who had had already bought their plots and who, backed by the *hacendado*, the *mayordomos*, and armed guards, had tried to defend their freshly conquered lands. The *rayados* were *arrendatarios* who wanted to take possession of their plots by force and were newcomers who saw the possibility to acquire land before the state could intervene and expropriate the property. The result was the end of La Granja's *hacienda* system and its spatial fragmentation into dozens of small farming areas owned by independent rural families (see Figure 3.3). The new owners began to build their houses with more solid materials (mainly adobe and wood), generally next to their farming plots. However, it was also the beginning of the formation of the town in the area where the *casa hacienda* used to be located.

What followed was the formation of a small rural town. Poorly connected with major regional markets and with small-scale agriculture for self-consumption, La Granja was a space that barely grew. Because of the limited size of their plots—making subdivision difficult—and scarce job opportunities, a significant proportion of the youth had little choice but to migrate to cities on the coast (Chiclayo or Trujillo) or in the Amazon basin (Nueva Cajamarca, Bagua, or Tarapoto).

In the early 1980s, however, mining exploration began in the area and a combined effort between Centromin, the Peruvian state mining company, and the German company, Sondi, identified a major copper deposit. Subsequently, Sondi constructed a mining camp and the narrow unpaved road, which today still connects La Granja with Querocoto, the capital of the district. Mining works were abandoned until 1994, when, through a privatization process, the central government granted a Canadian company, Cambior, a five-year option to explore and develop the project. Cambior initiated new exploration activities that provided jobs and income opportunities for the local population. At the same time, the company began purchasing land and implemented a poorly planned resettlement process. Using different strategies, including legal and physical threats, Cambior pressured local families to sell their land and displaced them into different areas. The central government, under the authoritarian regime of Alberto Fujimori, was eager to attract foreign investment and demonstrate the pacification of the country after the political and social violence and unrest caused by the Maoist group, Shining Path, a decade before. Thus, the central government closed local schools and health centers as a way to force the population to move from the area. In addition, government officials accused many of the local leaders opposed to the resettlement process of terrorism and initiated judiciary processes against them. As a result, of approximately 250 families that lived in the area 150 sold their land and migrated, mainly to Lambayeque (see Figure 3.4). In the village of La Granja itself, only seven families refused to move, and the company demolished the remaining houses.

This period marked the diaspora of *Granjinos* and converted La Granja into a derelict space. Among many *Granjinos*, the displacement was experienced as the end of a way of life they once knew: productive activities, labor conditions, social relations, spatial mobility, social landscape, sense of place, trust and vicinity were all disrupted. Without any support or an assistance program from the company, many the displaced families struggled to survive in the new setting and their living conditions dramatically declined. Confronted with new and challenging social, cultural, economic, and physical conditions, some *Granjinos* began to develop a sense of nostalgia and constructed an idealized and bucolic image of the La Granja of the past.

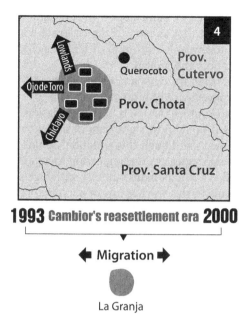

Figure 3.4 The beginning of mining-driven transformations.
Source: Castillo (2015, p. 69).

The displaced *Granjinos* constructed a rural *arcadia* where La Granja appears as a green space of abundance and social safety.

Since the 1990s, changes have accelerated. In 2000, Cambior sold its mining and surface rights to Billiton, which later merged with BHP. After technical and financial studies, the Anglo-Australian multi-national company concluded that the project was not feasible. Not without resistance and intense internal and external consultations and discussions, BHP Billiton decided to implement a program to remedy the environmental and social legacies left by the previous company.

At the internal level, local managers on the frontline had to over-come the reluctance of the company's central management in London to invest in such a program. The local managers made the case that activist groups against the company could use the situation of the dis-placed families to cause major reputational damage to BHP Billiton. In that time, the company was finishing a long and sensitive agree-ment with local communities in Papua New Guinea as consequence of its operations at the Ok Tedi Mine and management decided that the risk of becoming involved in another major controversy was too great.[2] The acceptance to implement a remediation action plan was facilitated by the fact that as part of the contract with the state, the company agreed to invest a minimum of US$ 3 million regardless of its final decision about the project.

On the external front, the company had to persuade state functionaries of the social convenience of returning the land to the affected families. The central government feared that the devolution of land rights to the local population would pose significant obstacles to future attempts to develop the project. Although this situation would prove to be true from a pro-business perspective, it is important to note that, de facto, securing surface rights would be a difficult task with high social and political costs. That would be the case because of population pressure over the land (especially from poor rural families),[3] general discontent, and lack of legitimacy of the previous displacement process, and state agencies with weak capacities and little political willingness to enforce the law.

Beyond the corporate strategies and decisions implemented regarding the project, what is important for our purposes is that the presence of a major mineral deposit alters the intensity and the quality of transformations, negotiations, and resistance over the space. In the context of increasing spatial fluidity (in terms of people and capital flows), mineral development introduces a fixity that will permanently shape the political ecologies of the subsoil in the region (Bebbington & Bury 2013a).

As part of its social program, the company offered the land back to the displaced families. Therefore, many families returned to La Granja (see Figure 3.5) and bought the land under quite advantageous conditions. However, this situation created an enduring conflict over the space between those who stayed in the area and did not sell their land to Cambior and those who returned. The first claimed that La Granja survived because of their struggle and, thus, that they should reap more benefits. The tension that arose between the resistant families and the returnee ones was especially bitter concerning the ownership of the lots in the urbanized part of La Granja and still shapes much of the local politics in the village. Trapped amid local pressures, the company transferred the urban terrain's titles to the Asociación del Señor de los Milagros (Lord of the Miracles Association), which, in turn, would distribute the properties.[4] Dominated by the local families, the association agreed to give priority to the children of the resistant families in the allocation of urban properties. Therefore, the association transformed from a religious organization into an important player in the local control for land access.

According to the estimates of one of the company's most experienced employees, around 75% of the families that sold their land decided to buy their land back. The new price that BHP Billiton established was less than

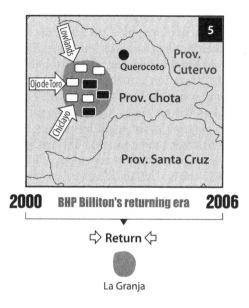

Figure 3.5 The returning process to La Granja.
Source: Castillo (2015, p. 71).

half of the original amount that people received from Cambior. In addition, BHP Billiton designed a credit system without an initial purchasing quota and with a period of ten years to pay (the first five years without interest and the remaining five at low interest rates). The funds raised would be deposited into a social fund—Fundación Paltic—for the development of the affected localities. However, many families accused the president of the foundation of misappropriation of the resources. After the president moved out of the region, most of the families stopped paying their debts. Nowadays, these families possess the land but do not yet possess legal property rights. The Rio Tinto La Granja (RTLG) project started a program to aid these families with the purpose of clearing the path in the case of an eventual land purchasing and resettlement process.

Of the total number of families that bought back land, between 20% and 30% returned to the area as complete households, between 50% and 60% had only some members return (for instance, the elders, or some of the sons or daughters after the inheritance was divided up), and the remainder regained ownership of the property but did not return to the area. These families received logistical support to return and initial capital in materials and farming seeds with the purpose of helping them to restart their lives.

The approximately 25% of families that did not choose to buy their land back requested compensation in cash. BHP Billiton agreed to give them a cash payment of roughly US$ 3,000, the monetary equivalent of what the other families had received in materials.

In 2006, Rio Tinto obtained mining rights from the central government and initiated exploration activities. As Cambior had before, Rio Tinto established programs for local employment. The relatively well-paid jobs (with salaries three times higher than those offered in the farming sector), opportunities for small businesses, and expectations for negotiating resettlement compensation and the sale of land to the mining company have prompted an enormous influx of *returnees*— relatives of *Granjinos* who had left the locality along different migratory processes. After years of decline, La Granja has become a burgeoning space of attraction and its Sunday market competes with the one in the district capital as the main trading center for surrounding villages.

In addition, families in La Granja are building new houses, expanding their old ones, and requesting social and recreational infrastructure for the village. This is a strategy that was implemented with the purpose of obtaining better compensation packages if the resettlement process takes place. With the implementation of the copper project and the construction of the open pit and other facilities, that infrastructure would be destroyed in another example of what Joseph Schumpeter (2010) and David Harvey (1999) call *creative destruction*. The original concept refers to the ability and need of capitalism to tear down everything it creates, in order to replace these with new products, buildings, and spaces at an increasingly frenzied pace. With his well-known assertion that "all that is solid melts into air"—a metaphor extracted from *The Communist Manifesto* of Marx and Engels—Marshal Berman (1982) emphasizes the condition and sense of fragility and evanescence present in a capitalist society. This ephemeral condition of capitalism imprints its hallmark in the materiality of society and its people's experience.

To date, employment, small businesses, and other benefit programs have been negotiated between the mining company and La Granja's leaders, who generally belong to the same family and live in the central part of the village. Families living on the outskirts of the village who also want to negotiate directly with the company have questioned this process. Consequently, families from different sectors outside the center are creating autonomous localities with their own authorities. This is the case of the newly formed villages of La Lima, Checos, and La Uñiga (see Figure 1.1). In this way, the physical and social space of La Granja continues a process of fragmentation (see Figure 3.6). Mining development is accelerating and deepening a phenomenon started many decades ago.

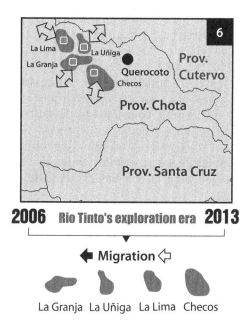

Figure 3.6 Current spatial connections.
Source: Castillo (2015, p. 75).

The described process of La Granja exemplifies how global flows and external trends are increasingly becoming more prominent in the configuration of different mining spaces to the advantage of local and regional forces. Since the 1990s, the mining development, with all its effects on land ownership and people's lives, has been shaped by liberalizing policies, global financial crisis (first in 1998 and then in 2008), and periods of international economic growth and high demand for commodities. These trends have led to the construction of roads, public infrastructure, mining camps, and houses, as well as the establishment of temporary jobs, the opening of small businesses, and—in general—the creation of a burgeoning place. However, as the previous shutdowns of the project have made clear, the link between global flows and mineral cycles has also produced a sense of this being an ephemeral place, which has generated stress, anxiety, and uncertainty in the lives of local people. As one male local villager expressed:

> ...I just want Rio Tinto to come and give me a price for my land... to give me an option... that [Rio Tinto] tells me "here is

the money and the new place where you will go". I don't want them to tell me that they don't know if they are going to stay or go. I want them to tell me something and I will take it or leave it for my family.

As depicted in Figure 3.7, global trends in mining development form part of a larger series of social and spatial transformations for La Granja and they significantly contribute to the destruction or transformation of what is being called a *sense of place* (Cresswell 2004). In this context, new, complex, and often contradictory senses of place—crossed by age, gender, wealth, family as well as individual experiences—are in a process of being elaborated, as the following chapters explore.

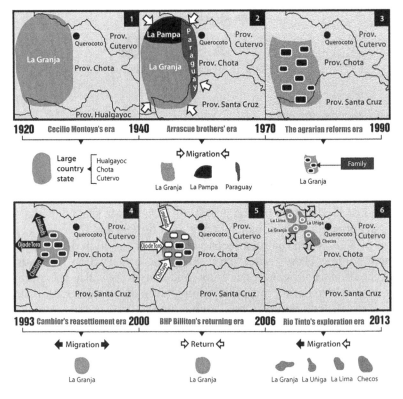

Figure 3.7 Spatial transformations of La Granja.
Source: Castillo (2015, p. 77).

Notes

1 Land scarcity, due to large-scale distribution among a few wealthy families, partly explains the pervasive rural poverty of the time. This concentration of land has its origins in colonial times but was expanded with the implementation of liberal policies after the country's independence (Deere 1990).

2 It is also worth noting that a growing concern around resettlement processes and the performance of mining companies and international lending institutions was emerging in those years. In Peru, for instance, researchers began to analyze the legacies of the land purchase and resettlement plans implemented around large-scale projects, such as Yanacocha and Antamina (Pasco-Font et al. 2001; Szablowski 2007).

3 Indeed, some displaced families were already returning to the area, illegally occupying their former land or clearing the forest for farming and housing.

4 The association was created before the arrival of Cambior and it owned and managed some pieces of land, which were sold to the company in order to cover its religious-related expenses. In addition to the urban terrains, the association received the nonclaimed farming lands, the local community center, and heavy machinery left by Cambior.

4   Access

In recent decades in Latin America, the control over territory has been at the center of social struggles and movements over access to natural resources and the defense of livelihoods (Bury & Kolff 2002; Bebbington 2007; Brain 2017). This has particularly been the case for mineral, oil and gas exploitation, foresting, and land claim for agro-export crops. Territory is a reality beyond its physical aspects and provides ideational resources (Offen 2003) in these struggles. Nor surprising then, territory has become the *locus* for the politics of identity of diverse groups in the region seeking higher levels of political, economic, and cultural autonomy (Hale 1997).

For around five centuries, land issues have dominated not only public policies and ideological and academic debates, but also many of the struggles over appropriation and resistance in Peru. La Granja, of course, is no exception. Access to land is central to the village's contemporary history and shapes much of its local politics and the identity of its people vis-à-vis cycles of mining development.

La Granja comprises around 1,200 hectares of land. Of the total, about 40% is natural pasture, 20% is composed of resting lands, farming lands occupy 15%, another 15% corresponds to unproductive lands, 5% is cultivated pasture, and another 5% is forest. According to the latest agrarian census, there were 125 farming producers in La Granja in 2012 and, taking the average of the whole district of Querocoto, around 70% owns between half and five hectares of land (Instituto Nacional de Estadística e Informática 2012a). The agrarian system in La Granja is composed of smallholders practicing subsistence farming with low levels of technical input.

The historical account of La Granja's space shows that land ownership in the area has changed hands on several occasions over the years, from one *hacendado* to another, from local families to a mining company, and then back to local families. However, the total amount of farming land is relatively fixed and hundreds of local families have already divided and appropriated that land. Indeed, forest clearance is one of the few ways to expand the area of farming land. The local forest is a natural protected area and the state forbids its cutting. However, many families undertake slash-and-burn practices in order to obtain wood and, more importantly, to access new farming land. Later, these families will seek the formalization of land titles from state agencies.

Of course, the destruction of the forest has its limits and the total area cannot be further expanded. In a context where it is difficult to expand the available land which has already been allocated, the mechanisms and barriers of access are crucial to the political economy strategies of the local families.

The forms of land access in La Granja have evolved over time but have also changed in terms of the actor who—legally and in practice—controls the access. The state and the market are the privileged space producers in modern, urban, and industrial societies. However, in cases of the fragile state and hybrid political order that the Peruvian society exemplifies, kinship networks and local social organizations play an important, though informal, role in facilitating or inhibiting mechanisms of spatial access. Using political, market, and kinship mechanisms, different actors have become involved in the complex and changing processes of land access in La Granja's spatial history.

4.1 The old regime: the *hacendado*

Before the agrarian reform initiated in the late 1960s, the whole land of La Granja belonged to one person, the *hacendado*. The *hacendado* then distributed small parcels within his property, generally in marginal sectors, to farmers while he kept the best areas for his own farming and for pastures for his cattle. These farmers, known as *arrendatarios*, paid the *hacendado* in cash, goods, and personal services. Amid the social and political turbulence that ultimately led to the reform, the local *hacendados* began to distribute part of the marginal lands of their properties.

Thus, for instance, the father of the male informant of the Resistant family 2 bought the farming land he was working from the *hacendado* and, for many years, paid for it with *yuntas* (two oxen that plow land with a hoist). In a different situation, the male informant of the Returnee family 2 mentions that after the agrarian reform, the state allocated vast farming terrain to the ex-*hacendado* in La Granja. However, when the Arrascue family moved out of the village, the *rayados* occupied the land. In this context, and because he was living in extreme poverty, the informant invaded part of the area. In the early 1990s, the mining company, Cambior, initiated a land-purchasing program which required the regularization of land titles. Thus, backed by the company, the ex-*hacendado* returned to the village to sell their land to the current tenants, who in turn were able to sell the titled land to Cambior. The informant paid one head of cattle for the land. In another case, the male informant of the Migrant family 1 mentioned that at the beginning he worked on leased parcels, but that he was later able to buy 17 hectares of land in the La Lima sector:

> I am from Olmos [in the Cajamarca Region]. I came to the area because the Arrascue brothers were my cousins. I paid for the land with a *yunta*. I bought my whole land little by little, given the *hacendado* my cattle. Later I got the public deed.

It is interesting to note the existing family networks between *hacendados* and *campesinos*. Indeed, both actors used kinship relations to build a system for exchanging goods, services, and favors in the absence of market mechanisms. The system was unequal and oppressive, but it provided a shared world of cultural values and a natural order, with the patriarchal family as an ideal model. In this paternalistic world, the *hacendado* represented a father who sought the best for the *campesinos*, his children.[1] It is unsurprising, then, to find among villagers' feelings and memories of gratitude and loyalty toward the *hacendado*.

The case of an old villager from La Pampa reveals many of these features. The old man was born in Cutervo, a province of the Cajamarca Region, and his parents migrated to the La Granja *hacienda* to become *arrendatarios* of the former *hacendado*, Cecilio Montoya. Soon after the family migrated, he became an orphan at a very early age and began to work for the *hacendado*. When the owners of the *hacienda* changed, he became an *arrendatario* for one of the Arrascue brothers. He was one of the few people who defended the *hacendado* during the struggles of the agrarian reform. In the area, two factions were formed: the *amarillos*, who were renters that had already bought land and supported the *hacendado*, and the *rayados*, who promoted the occupation of land. In this confrontation, the *rayados* killed the *hacendado*'s cattle or took them to other places. Several times, he went to take the cattle back and despite the danger and tension, nobody attacked him. He said, "I am not afraid, the Lord will decide". During this period of social convulsion, the *hacendado* Alejandro Arrascue proposed to sell him part of the land he was farming. The villager selected a good place next to the Paltic River and agreed to make monthly payments. However, after the second installment, the state granted him the land title. The farmer showed the title to the *hacendado* and told him that he had not taken any steps to receive the deed. Alejandro Arrascue replied, "You have given me good support, now I will help you". Thus, the *hacendado* did not accept any other payment and gave him the title. For the old villager, the *hacendado* represented a good person for whom he had happily risked his life.

Many of these changes have occurred over the span of one generation and many *campesinos* have shifted from being *arrendatarios* to owners and are, therefore, conscious of the experience of social mobility.

4.2 The state

Many local families received land from the state during the agrarian reform. Therefore, in the collective memory of many *Granjinos*, especially among the elders, social identity is tied to the shift in the socioeconomic and legal property regime—from the *hacienda* to the smallholding system, when La Granja's families were recognized as owners of the land after their struggles with the *hacendado*. The opening paragraph of a program of the main religious festivity in the locality vividly captures this collective memory[2]: "[In the day of] ...the *Granjina* identity, we honor the fight for the independence from the *hacendado* yoke as a way to rescue customs from our ancestors and strengthen them...".

In *Añoranzas de mi tierra* (Longings for My Homeland), the poetry book of local writer José Guevara, there are some verses that depict this period of struggles over accessing the land in La Granja. For instance, in the poem "Los que llegaron" (Those Who Arrived) Guevara remarks:

> Arrived landlords, imposing
> Rules that favored the master,
> They gave the land on lease
> Each *Granjino* became a miserable peon.
> And at the brightness of so much exploitation
> The agrarian reform arrived,
> They parceled the land
> And each one cultivated the land.

The poem "Hitos relevantes de mi Granja" (Relevant Milestones of My Granja) highlights even more the tension between an exploitative space and a space of liberation:

Powerful landlords
Imposed their arrogance,
The Montoya and the Arrascue
That raise their earnings.
Humble and Andean family
Worked in its leased land,
And two days per week
Were devoted to the landlord.
The agrarian reform attempts
To give land to the *campesino*,
Around the 1960s
This great good fortune came to us.
The battle had a conquest
At the vanguard was a *Granjino*,
Godofredo Guevara was
At the front of the *campesino*.[3]
The fields were cultivated
With dedication and humility
New life began
With pride and freedom.
With own and modest houses
The village began to grow
Never despicable masters
La Granja has had again.

These lines express the agency and will of the *Granjinos* to make the land reform happen. It is true that the state intervened, and that this was perceived as being a "great good fortune". Nevertheless, the *Granjinos* had to fight to conquer the land. Moreover, we must note that this is the only moment when the state explicitly took part in La Granja's history according to this long poem of relevant milestones of the village. Indeed, the state is perceived, in the best case, as absent in the zone and, at worst, as actively supporting foreign and private interests. Hence, during Alberto Fujimori's regime, the central government closed the public schools and the medical centers of La Pampa, La Iraca, and La Granja as a measure to press local people to sell their land to Cambior.

In addition, the central government opposed the land return program proposed years later by BHP Billiton. Nowadays, the government enthusiastically seeks to title local families' land in order to facilitate the development of the mining project by Rio Tinto. To be sure, since the early 1990s, the influence of liberalizing policies and neoliberal discourse in Peru has been hegemonic. In 2007, the then president, Alan García, published a series of three opinion columns in a major conservative newspaper in the country under the title, "The Dog in the Manger Syndrome". In his opinion, a large extension of land—as in the case of other natural resources—is not a tradable commodity, does not receive investment, and does not generate employment in the country. The consequence is the persistence of poverty: "And all because the taboo of superseded ideologies, idleness, indolence, or the orchard dog law, which says: 'if I not do it, that nobody do it then'" (García 2007, p. 1).[4]

With the state playing an ambivalent and somehow contradictory role, which swings between eagerly promoting private investment and fulfilling its duties to warrant citizens' rights, it is not surprising that the most used and trusted links in order to gain access to land are the family networks.

4.3 Mining companies

Only after the agrarian reform, properly land purchases among individual and free owners began in the area. Nevertheless, because of the small size of the plots and poor farming outputs, transactions were scarce. It is with the arrival of mining development in the zone that land began to increase in value and the acquisitions increased.

The first land sales took place with Cambior. After winning the option to explore and develop the copper project in 1994, the Canadian company began an aggressive program of land acquisition. Many of the land transactions go back to these years. The male informant of the Resistant family 2 narrates that his father owned 60 hectares of land and that Cambior had made an offer on them. However, before signing the contract, the male informant intervened and negotiated a significantly higher price with the company representatives—200% higher than the original offer. After the sale, the male informant accompanied his father to buy land and settle in Chepén, a small coastal town in a fertile and productive valley. His father began to cultivate sugarcane and rice and raise livestock. This is the case of a relatively successful migrant during the diaspora created by Cambior. One of the reasons for his success was the initial large extension of land sold and the decision to invest in a profitable coastal farming area, which is well connected to main cities. In addition, the advice and support of his son made a substantial difference. When Cambior arrived in the area, the male informant only owned a house in La Ganja. In 1995, his father gave him and his younger son two hectares of land each as inheritance. His father transferred these properties with the aim that his sons would be able to negotiate a better compensation agreement with the company. However, when the informant returned from Chepén, the company did not make an offer for his land:

> [The project representatives] ...told me that I had already secured a good price with the sale of my father's land and that they had no more money to buy my property.

The male informant was never able to sell his property to Cambior and by default he became resistant to the mining project.

Many other families, however, sold their land under the pressure exerted by Cambior and suffered very poor outcomes. This was the case of the Migrant family 2. The male informant sold the land inherited from his father and bought a house in a peripheral neighborhood of Chiclayo and some farming land in the province of Ferreñafe, two hours by bus from the city.

In 2001, after conducting technical and financial feasibility studies, BHP Billiton found the project to be unviable and implemented environmental and social closure plans. As explained previously, many families decided to buy back the land from the company, although not all of them returned to the village to live. This period was also a time of complex arrangements in land ownership among local families. For instance, the father of the male informant of the Resistant family 2 purchased the land from BHP Billiton through one of his sons and, later, the son transferred the terrain to his father. At that time, the company offered two options: the opportunity to buy back the land at a reduced price with soft loan conditions or to receive one single payment of around US$3,000. Therefore, using his son as an intermediary, the male informant's father obtained both the land and the cash bonus. His father now hires workers to cultivate the land and enjoys a relatively easy life.

The male informant of the Migrant family 1 explains that he and his wife did not want to buy their former land and return to La Granja despite the insistence of his sons and daughter. Finally, the children purchased the 17 hectares of farming land, which is currently being leased to cultivate pastures for cattle raising. This case illustrates the fact that not all the families were interested in moving back to their former villages. This was the situation of the male informant of the

Migrant family 2, who sold the land he had recovered from BHP Billiton, unaware of the arrival of Rio Tinto years later. Nowadays, he has started to buy land in the area. In addition, the male informant of the Migrant family 4 explains that during the displacement process of the 1990s, his father sold his three-hectare terrain to Cambior. When BHP Billiton implemented the social closure plan, he wanted to purchase that land. However, he had no money and his siblings were not interested in the investment. The male informant has since bought some small plots in El Verde, a nearby village, and his eldest son works the land alternating with temporary jobs at the mining camp. Many of the decisions about whether or not to return to the villages of origin have been influenced by a combination of access to capital to invest in the land purchase, social and economic situation of the family in the resettled area, and future expectations. Indeed, expectations of future economic benefits shape many of the strategies concerning how the local families perform with the mining companies.

4.4 Local families

In anthropological literature, kinship is one of the most important webs of social relations that organize human lives in many societies (Rivers 1968). Kinship systems include people related by both descent and marriage. That is to say, kinship integrates members of one ancestry or lineage—that could be patrilineal, matrilineal, or bilateral—and extends this network to other groups through a marriage alliance. For Andean societies, the use of extended family networks is one of the most common and important strategies to access, produce, allocate, and consume goods and services (Mayer & Bolton 1980; Ossio 1981). Land is not the exception and it is transmitted within the same lineage through a descent system; or, it may be obtained from another group through marital alliances.

In La Granja, people use a combination of both systems. Hence, for instance, the male informant of the Returnee family 1 inherited some farming land from his father. In 1994, Cambior was pushing local families to sell their land. The informant's father decided to sell but, beforehand, he divided the land, kept one parcel for himself and distributed the rest among his children. With this strategy, each child was able to sell their parcel to the mining company and move out of La Granja to live with their own family. In 2002, amid the returning process promoted by BHP Billiton, the male informant bought back the parcels of land that had belonged to him and his siblings. Nowadays, he rents the whole land to his father, who uses it to cultivate pastures for cattle.

In the case of the male informant of the Returnee family 2, he divided his ten hectares of land among his eight children and in 2011 gave it to them as inheritance. The children will be then able to build houses in that terrain to negotiate with the mining project for compensation benefits.

Inheriting land is not only employed as a mechanism to negotiate with the mining company. Local people use this mechanism for different purposes. The female informant of the Opportunistic family 1, for example, inherited a piece of farming land in the neighboring district of Querecotillo from her parents. Years later, she sold half of the land and gave the money to her first daughter, who had suffered an automobile accident. She then gave the other half of the land to the same daughter, who moved with her family to live there. This family then divided their house in La Granja into two sections and gave one part to their fourth daughter, who, after separating from her husband, moved to the village with her daughter in 2008. Thereby, the land transfer was used to support four generations.

Hence, inheritance provides a flexible structure that people employ to create complex arrangements among the members of the family network, depending on specific circumstances. It is an asset that can be used in moments of crisis or for taking advantage of opportunities when they arise.

Marital alliance is the other strategy that allows people to enhance family networks and to access goods and services beyond their own original kin. Thus, outsiders—such as schoolteachers or nurses—can access land by marrying local men or women. For instance, the male informant of the Opportunistic family 2, a former schoolteacher from Chota, in the Cajamarca Region, moved to La Pampa in 2006; and, after living in the village for two years, he became a member of the *ronda campesina*. As part of the *ronda* he was able to secure employment at the temporary program established by RTLG. However, it was only after he married a local woman that he was able to access land. His mother-in-law divided and sold a large piece of land among her male children and the male informant with the intention that each child would eventually be able to negotiate with the mining project.

Indeed, in the context of an immature land-market mechanism, newcomers find a way to access land through kinship networks. Hence, *Granjinos* who moved out of the area, or relatives who were born in other localities, could return to the village and claim land rights. Nevertheless, along with kinship relations, people have used monetary transactions since the years of agrarian reform. Moreover, the growing interest for land and the arrival of a significant number of newcomers in the area now limits the use of kinship as a mechanism to control land access and land distribution and to better exploit economic benefits. Consequently, local people are increasingly using market mechanisms to access and allocate key resources and to maximize their individual and family benefits. To be sure, many of the land transactions are arranged within family networks. For example, in 2001, the male informant of the Returnee family 1 bought some terrain in the urban sector of La Granja from a cousin who had not sold his land to Cambior. The same year, the male informant built a house. This was the first house the family had occupied after returning to the village from Chiclayo. Two years later, the informant built a second house, where he and his family moved and in which they currently reside. From 2002 to 2005, he handed over the house to his cousin who was in charge of maintenance and the payment of utilities. In 2005, the informant leased the house to his father and two of his sisters until 2012, when he began to rent the place to an outsider family.

Land transactions in La Granja, of course, extend beyond own family networks. This situation is becoming more common when the interest in land evolves more dynamically. Generally, a market becomes more dynamic because the number of players increases, the players' wealth grows, or a combination of both factors. Both elements are present in La Granja; there are more newcomers—relatives of *Granjino* residents in the village or outsiders without local family networks—willing to access land and the local families enjoy greater purchasing power.

Some of these transactions were made before the arrival of mining companies in the zone. For example, the Resistant family 3 bought a piece of land from a local family when they arrived in the village in 1992. The following year the family built their house in the terrain and started to rent rooms for schoolteachers, who were mainly from Chota.

Other interviewed families bought land from local villagers during some of the mining downturns. The Returnee family 1 bought two pieces of farming land along the road that connects La Granja with La Pampa in 2000. They purchased two hectares from one family and four from another, which had already been bought back from BHP Billiton; however, this family was not interested in returning to the region. In that time, nobody thought that a new mining company would arrive until years later. As the male informant of the Migrant family 2 explains:

> In 2000 I bought back my original farming land from BHP Billiton but then I sold it to other people. I did not know that another company would come to purchase the land again; otherwise I would have kept it. Now, I have bought an urban plot in La Pampa through the *ronda campesina*.

Until 2006, pressure over land was not excessive. The arrival of Rio Tinto, however, changed the situation and local families are increasingly seeking new land and properties.

The Opportunistic family 2 has bought three plots for housing purposes over some years. In 2007, they bought a plot in the urban sector of La Granja from a local family and built their house. In 2009, they bought a second terrain in an area close to the mining camp. The plot was part of a bigger land property that belonged to the male informant's brother-in-law, who subdivided the land and sold the plots to newcomers interested in building houses with the intention of eventually selling them to the mining company. In 2011, the family bought a plot located next to their house and expanded the property. It is interesting to note that the male informant redirects the income he obtains from working in the RTLG's employment program into the acquisition of other properties, with the ultimate objective of selling these properties to the company.

These cases illustrate the way in which some local families can mobilize and direct resources obtained from their strategic position in the control of the space—and thus benefiting from the development of mining activities—into the acquisition of more land and properties in the area surrounding the mining project or in the neighboring villages. It is also interesting to note the extent of inequality and power imbalance the temporary land acquisition and ongoing land renting of the company has created.

One of the strategies that the examined families have developed is to acquire as much land as possible and distribute it among their family members with a double purpose: as a unit, the family receives a higher monetary compensation, and, individually, each member can share part of those benefits. In consequence, the land market in La Granja is turning into a more active one because the exchange value of the land has sharply risen. With the exemption of current urban development in the village, this increase has little to do with the rise of current land-use value. Indeed, the use of land for farming purposes has declined. Land prices are rising, owing to the dynamics of land speculation resulting from the prospect of the copper project development.

Another common strategy that local families employ to increase the benefits they can possibly obtain is to improve current houses or purchase additional urban property. Certainly, all the interviewed families currently living in the area have used variations of this strategy. They embark on these investments seeking immediate returns (for instance, renting rooms to newcomers) or expecting higher future benefits in the event of a negotiation agreement with the mining company.

In the latter scenario, the families will have made improvements to their current property; for example, building additional levels and replacing old materials such as mud, wood, or straw with modern ones such as brick, concrete, and glass.

Although the provision of housing to outsiders has been one of the family's economic strategies for a long time, the arrival of mining companies has substantially amplified the dynamics. In this line, the Resistant family 3 represents one of the most notable examples. The family derives its wealth not from farming activities but from the provision of services linked to the mining development in the area. The couple runs a small grocery store and a restaurant in the village. The second son works for an engineering company that provides services to mining operations in Cajamarca City, and, during his time off work, he hires a truck to transport food between La Granja and Chiclayo. The third son works for Sodexo, a subcontractor of RTLG that offers the catering service to the camp. Their daughter's partner is a bus driver who works for a family-owned business that covers the route between La Granja and Chiclayo. In addition, one of their most important sources of income is the lease of rooms to schoolteachers, health workers, and, more recently, to employees of consulting companies linked to the mining project. For this reason, it is critical for the family to acquire terrains and houses in the urban sector of the village.

Nowadays, in addition to some farming land in La Granja, the family owns four houses in the village: three urban plots in La Pampa and one terrain in Picsi, a town near Chiclayo.[5] In 1991, when they arrived in the village, the family lived for two years in a house leased from a female informant's cousin. In 1992, the couple bought a terrain from the female informant's father. They paid in a combination of cash and farming products[6] and the next year built their first house. With some improvements made in 1998, the house now boasts three levels and fourteen rooms, which the family rents out. In 1994, the female informant inherited a terrain near the Checos River from her father that was used to raise pigs.[7] On that terrain, they built a second house comprising two levels and four rooms. Currently, the family rents the first level to an engineer, a relative of the female informant, and the third son and his partner use the second. In 1997, the female informant received some inheritance from her father, a piece of farming land in Checos. In 2005, the couple acquired a terrain, next to their second house along the Checos River. They bought the terrain from a female informant's sister and although the couple is the owner, the title is in name of their first son. In the terrain, the family built a third house between 2008 and 2012. It has better internal finishing, which is part of a strategy to obtain a better price from the company if the resettlement process goes ahead. The house has three levels and six rooms, which they lease to senior personnel of consultancy firms working on the mining project. In future, the couple plan to open a restaurant in the house. This plan seems to be consistent with a strategy to try to obtain the most possible benefits from mining activities in the locality without selling the land and moving out of the area. Indeed, at least discursively, the couple is among the main opponents to the development of the mining project. However, this opposition should be understood as a strategy that assists the couple in securing benefits from the mining project without losing land rights.

In addition, in 2007, the couple bought three adjacent plots in the nearby village of La Pampa. The terrains belonged to the female informant's relatives, who were in a health emergency, and the couple made cash payments for the purchase. They registered the plots in the names of their children. The male informant was advised that for negotiating with the mining company, instead of concentrating properties in the name of one person, it would be better to distribute them among different adults, who could then try to be considered as different families. Finally, in 2012, the couple bought a fourth house in La Granja that leased it to a mining worker.

The trajectory of land sale, acquisition, and division of the later family shows a complex strategy based on investing the income generated through mining-related activities—direct employment in temporary job programs offered by the company or small businesses supplying goods and services to the company and its subcontractors—into more land, urban plots, or new houses in the area. The objective is far from increasing farming production or satisfying housing needs of the family members. Instead, the aim is to enhance the opportunities for a better compensatory negotiation with the mining company in the event of a resettlement process.

Through the analysis of these families' stories, it may be observed that the increasing importance of housing and urban plots is a significant factor regarding transactions for accessing land in La Granja. It is worth remembering that until the 1970s the manor house was the only major building in what is now the urban sector of the village. The farmers were not allowed to build houses of brick and concrete; the urban landscape was composed of small and scattered huts next to the farming plots. After the agrarian reform, some of the newly free farmers began to build their houses around the manor house—which was destroyed during the struggles—initiating the creation of the urban sector of the village. In the aftermath of the displacement caused by Cambior, La Granja was transformed into a derelict space. As one local villager remembers, "when I left the place in 1998, only seven houses remained in the village, the others were destroyed, it was very sad to see La Granja". The returning process promoted by BHP Billiton in 2001 initiated the resurgence of the small village. Nevertheless, most of the houses were still located next to the farming areas of each family. This situation has dramatically changed within the context of recent mining development; the number of houses in La Granja has grown from 7 or 9 in 1994, to around 30 in 2004, and between 140 and 180 in 2013.

The general urbanization process of the country, and the need for its inhabitants to group in more dense settlements in order to access public services, such as electricity, partly explains the rapid growth of the village.[8] The local expectation to negotiate higher compensation packages in case of a new resettlement process is, however, the main reason for the frantic construction of new houses.

A related process is the acquisition of land and houses outside the region. In this sense, it is symptomatic that six of the seven interviewed families living in La Granja have another property in Chiclayo. For some families, the property they own in the city is considered their main residence and the place where they would go if the mining project is developed. The male informant of the Returnee family 1 explains:

> We bought a house in Chiclayo in 1996, during the resettlement process of Cambior. We went there thinking of the education of our children and because as a merchant, I established my working center there. My two oldest boys live there although my family live alternately in both places. Now, our main residence is here in La Granja, because the business is good and because our children can stay here during their holidays. In the future, our main residence would be in Chiclayo though. Everything depends on the negotiations with the mining company.

This phenomenon has created a patron of double residence: families with properties in the city but who do not lose control of their assets in the country. This pattern produces and is maintained through complex and seasonal flows of people, goods, and money between urban and rural areas. Decisions about who goes where, when, and for what reasons depend on family calculations in terms of job opportunities, economic benefits, or future education goals, which are in turn underpinned by age and gender considerations. Far from static rural populations, what we observe is a fluid landscape that connects the country with the city. By means of the use of diverse spatial practices, such as kinship or gender relations, local families blur the boundary between city/country and create more comprehensive spaces to live.

Notes

1 In many parts of the Andes, the *campesinos* called the *hacendado*—or any powerful actor—"tayta", which in Quechua means "father".

2 The *Señor de los Milagros* is the major saint and patron of the village. Every October a festival is celebrated in his honor, and it is the most important celebration in La Granja.

3 Godofredo Guevara was the historical leader of La Granja in the struggles for land. However, he has later been accused of illegally selling the terrain of the school to Cambior and was expelled from the village. The author of the poem is the nephew of Godofredo Guevara. It is clear that he attempts to depict the family as being heroic in this narrative of the village's saga. The landscape is read in kinship code (Rumsey & Weiner 2004) and the narrative serves the local "politics of place" (Moore 1998).

4 Bebbington (2009) argues that García's statement is an expression of a broader push in Latin America "to open up frontiers for extracting hydrocarbons, mining, producing biofuels, harvesting timber, and investing in agroindustry" (p. 13). What is surprising is that this expansion was occurring in vastly different political regimes, from Peruvian and Colombian neoliberal ones to Brazilian, Ecuadorian, and Bolivian, allegedly progressive, regimes. One of the many reasons for this convergence is the need for resources to finance social programs that are essential for the functioning of those regimes (Bebbington & Humphreys Bebbington 2011).

5 The family has an additional terrain in La Granja. When BHP Billiton opened the urban plots of the village for sale, one of the children sent a request to the Asociación Señor de los Milagros for a terrain that was used as garbage dump. Although the association rejected the application, the family has fenced and filled the area. Because the terrain is involved in a legal dispute, the family does not consider it a part of their property yet.

6 Four sacks of rice and approximately US$ 30.

7 The female informant indicated that she received this terrain as inheritance from his father. However, in order to avoid distributing his other properties among the rest of his siblings, the father gave her a deed of sale before a notary public.

8 The state, with support of RTLG, provided with electricity the villages of the area in 2008.

5   Production

Some of the literature from the social sciences on the Andes has depicted a social and economic landscape that equates rural societies with agrarian ones. Many Andean cultures, of course, have been great agrarian civilizations and, to be sure, the Inca State controlled a vast territory and a large population through the appropriation and distribution of agricultural surplus for its own political purposes. However, it would be misleading to consider that current rural Andean populations exclusively, or even mainly, dedicate to agrarian activities. Indeed, since early colonial years, rural populations have actively engaged in activities outside agriculture. The flow of workers to urban and mining centers (Assadourian 1982)—at various compulsory levels— or the blossoming of *obrajes* or garment textile workshops (Salas 1998) are just two samples of these activities.

In addition, as any other, rural Andean populations act within the possibilities and constrains of political, economic, and historical forces. Hence, rural households' decisions regarding the allocation of labor are taken within broader integration between international markets and regional economies. In Peru, for the last five centuries, the particularities of economic integration have been strongly linked to commodity-export booms: silver, sugarcane, cotton, oil, copper, or gold (Thorp & Bertram 1978). Therefore, the shift from farming and nonfarming activities and the specific allocation of time for family members to engage in these activities is part of the households' decision to maximize and complement their benefits, and minimize their risks according to the resources (namely, land, labor, and capital) they hold, and the priorities they define in the context of external opportunities and limitations (Zoomers 1999). Consequently, rural households will tend to allocate more labor and time to nonfarming activities if their agrarian assets (land, water, forest, pastures, and technology) are too scarce to generate enough income for the survival of the group

(Velazco 1998), or when stronger market integration provides incentives to obtain higher income (Escobal 2001). In the case of the latter strategy, household members will return to farming activities when economic dynamism decreases.

Following this logic, Long and Roberts (1984) and DeWind (1987) interpret the formation of the figure of the peasant-miner as coinciding with the emergence of large-scale mining in the central Andes at the beginning of the 20th century. Typically, the peasant-miner was a male from the *comunidades campesinas*, surrounding the operations of the Cerro de Pasco Copper Corporation, who worked in the mining operations but returned to his community to cultivate the land during labor peaks of sowing and harvesting. Although peasant-miners developed a class-consciousness vis-à-vis their relationship with mining companies and state representatives (Flores Galindo 1993), they did not lose their *campesino* identity and land-based economy. Indeed, until the last century, there were very fluid communication channels between rural communities and mining locations, because of the need for labor, services, and goods (Contreras 1988) as well as the transference of capital between farming and mining sectors (Manrique 1987). However, the arrival of large-scale mining—with its capital-intensive and enclave features—disrupted previous regional circuits. Local communities suffered, then, from a lack of benefits and environmental damage (Helfgott 2013). The "new mining", nevertheless, exhibits different characteristics and presents different challenges in the country. Through its corporate responsibility programs and policies, which have designed local procurement and local investment programs, and through the system of income tax distribution, which has allocated millions of dollars to mining regions, mining development has unleashed extraordinary economic and political dynamics (Arellano-Yanguas 2011). In addition, the "mining town" model, epitomized by cities like La Oroya and Cerro de Pasco, has given way to new patterns, wherein large mining operations have not created ad hoc urban settlements but instead make use of nearby cities for their housing needs—as is the case of the Yanacocha and Antamina mines in the cities of Cajamarca and Huaraz, respectively (Vega-Centeno 2011).

Failure to recognize the complex interplay between farming and nonfarming activities and the rationale behind rural household decisions obscures an understanding of the degree to which particular patterns of mining development helps to shape regional dynamics of production. There follows an account of the changes in farming and nonfarming activities along the different stages of mining activity in La Granja.

5.1 Farming activities

La Granja is a rural area where economic life and collective identity are strongly associated with farming over time, from the *hacienda* era to the present day. To be sure, before the arrival of mining projects in the area, farming was the main activity, in terms of income and allocated labor, for the local families. Corn, potato, manioc, sweet potato, *arracacha*, sugarcane, beans, and pumpkin were the main crops. Because the families dedicated more time, effort, and labor to agriculture, the total production and productivity were higher than that of current times. Even for some informants, it was possible to obtain two harvests per year because some families used artisanal water channels to irrigate their lands. However, a large proportion of the crops depended on seasonal rains.

Unlike other Andean regions, in La Granja there is no collective land use or water use. As it has been previously noted, land is owned by individual families and there is no collective organization with authority over land access and land production. Nevertheless, much of the farming work, especially during the peak times of sowing and harvesting, was undertaken with the reciprocal help of relatives and neighbors. As a local male stated:

> In the past, [...] all the work was done through the "assistance with assistance" system, where you returned with labor the received assistance. The family that received the assistance provided food and drinks to everybody. For that reason, farming was profitable. The community assistance avoided the use of money to farm the plot.

Therefore, in La Granja, the use of a "symmetric reciprocal system" (Alberti & Mayer 1974) was less of an expression of a collective organization than a strategy to access manpower in a context of limited monetary resources. Local farming production was and still is labor-intensive and employed much of the population. A local woman describes the prevailing production arrangements of the time:

> All the people used to work on agriculture and cattle raising. When the family wanted to go to Chiclayo to visit and buy clothes, they sold an animal. The harvest was stored in a barn and then consumed throughout the year. The family surplus was exchanged with the neighbors for other farming products and thus the family obtained the products they did not cultivate.

Although most of the family's production was consumed within the same unit, a portion of the surpluses was exchanged in a symmetrical relation within the village, and some was sold in local and regional markets. Until the 1980s, Chongoyape, a crossroads on the route to Chiclayo, was the most important regional market. Over a three-day journey on foot, the *Granjinos* traveled there to sell their agriculture surpluses and cattle and buy other products such as salt, kerosene, detergent, etc.

Under this inefficient agrarian regime, and in the absence of public services and a welfare system, families and people with insufficient land lived in poverty; and many of them had little choice but to migrate, especially to the coast. This is the case of the female informant of the Migrant family 3. The female informant was the daughter of a family with scarce land who had 13 children. When she was 15 years of age, she went to work on the coffee plantations in Jaén, in the lowlands. After contracting yellow fewer, she moved alone to the outskirts of Chiclayo to work in the rice fields under extreme conditions: "it was very hard, to work in the water, I had colds constantly and had lower-back pain from bending down for many hours".

In brief, before the arrival of the first mining developments, a subsistence farming system prevailed in La Granja. Limited inputs, low productivity, low levels of connectivity, and high transportation costs characterized the system. In addition, the families used nonwage employment, dedicated the major part of their production for on-farm production, and sold the scant surplus to local markets. Compared with current farming practices, this productive system is thought of as being closer to nature, based on solidarity rather than profitability and selfish individualism, and in association with a patriarchal extended family model. In addition, agriculture involved the majority of the population and working time and was the axis of the local identity. The name of the place literally translates to "The Farm".

Nevertheless, in the late 1970s, mining works began in the area, which altered production patterns. The German company, Sondi, built a camp next to the La Granja village and, in 1982, constructed the dirt road that connects La Granja with the highway to Chiclayo. The road allowed the arrival of the first vehicles to the village. Because of the persistent conflicts between the *Granjinos* and the allies of the *hacendado*, the *Granjinos* were not allowed to trade in the market of Querocoto. Given these circumstances, they bought a truck with the purpose of directly trading their products in Chiclayo. At the same time, a cattle road that connects La Granja with Maichil—en route to Chiclayo—was built. The later road was shorter than any previous route to reach Chiclayo, and the families began to use it to send their children to school in the city and to sell their cattle.

The construction of these roads contributed to a shift in the local and regional centers. With a faster and more direct connection to Chiclayo, Chongoyape consequently declined as a regional market. In addition, despite some nearby villages, such as La Iraca, having more families, farming land, and greater agricultural production, La Granja began to situate itself as a local center because that is where the road ends and is the final stop for trucks and buses. La Granja's positioning as a collection center for agrarian production was reinforced when, in the early 1990s, the Sunday market was established in the village.

Nowadays, La Granja continues to be the local center for trade, even competing with Querocoto, the district capital. Nonetheless, from being a farming producer, La Granja has transformed into a consumer of farming products. Needless to say, the poorly planned resettlement scheme implemented by Cambior destroyed any farming production in the area. It was only after the return of some of the displaced families under BHP's social closure that farming activities began to recover. With the arrival of Rio Tinto in the area and the implementation of an exploration program, however, farming production has sharply declined again. Some explanations point toward the negative influence of environmental effects. Thus, a few informants note that the mining works executed by Cambior have affected agriculture, mainly because the entrance of heavy trucks into the farming plots has compacted the soil. Nevertheless, they indicate that after cultivating pastures, the soil is recovering to its original quality. Indeed, after a stabilizing process, agriculture regained previous production levels and, in some cases, even showed improvements thanks to the new techniques that some families brought back with them from their experience in coastal areas.

In a similar vein, there is a perception among several villagers that nowadays there is less rain, the temperature is warmer, there are more pests and plagues, and that the land produces less. And they believe that this situation is somehow caused by mining activities.

Current researches (Bury & Kolff 2002; Bebbington & Williams 2008; Himley 2014), examining the changes that mining activities bring into agrarian economies, have tended to focus on the deleterious impacts on the natural environment, especially over the quantity and quality of available water. Following this logic, mining activities would produce a deteriorated environment, which leads to declining farming production and, in turn, would prompt social conflict. In the

Cajamarca Region, where water is one of the most significant elements in the long-standing and highly visible conflict that opposes local population with Newmont, it is not surprising that environmental concerns around mining operations are ever-present.[1] Framed within a political ecology analysis, these researches have the virtue to highlight how environmental arguments become tools for social and political struggle. However, they have more limited value for the understanding of the causes of productive variations. As we state, an important part of the changes driven by mining activities is not necessarily related to environmental linkages and could have been initiated long before any significant physical operation had been conducted.

A household's decision to redirect the amount and intensity of labor from agriculture to nonfarming activities is the most significant factor in explaining the acute decline in agrarian production in La Granja's case. The arrival of Rio Tinto, and the implementation of local employment and local business programs as part of the company's social management strategy, created unprecedented opportunities for local families to substantially increase their income. For instance, while the daily wage in agriculture was US$ 2.5 in the temporary employment scheme it is US$ 12.5, roughly a difference of 400%. Under these circumstances, local families seek to invest most of their time in mining-related activities. Therefore, farming receives a considerably lower number of working hours. In addition, since adult men tend to receive many mining-related job opportunities,[2] there is an increase in the share of farming work for women and old people. A local woman explains, "Half of the *Granjino* families continue cultivating the plots but in a less intensive way. This is because farming workers ask for higher salaries, which is not possible to pay because then the farming production would not be profitable".

Farming production has drastically decreased. Farming is conducted using the labor within the family network and/or the hiring of a few workers. In their choices to cut farming production, families tend to limit the production of staples, such as potato, corn, or manioc, just to satisfy the needs of direct consumption within the household unit, while keeping the most profitable and market-oriented crops, such as granadilla or coffee, which have benefited from the improved transport conditions to Chiclayo. The other crops are not prioritized, mainly because they cannot compete in price with products from the coast; as a female villager argues: "there is less work in the plots because people have cash to buy products from the coast in the Sunday market; manioc, corn, beans, and sweet potato are coming from Chiclayo or Quipayuc".

In addition, the increase in local wages and the shift in the allocation of labor have gone hand in hand with changes in consumption patterns. Therefore, many local families have increased their overall food consumption, opted for better-quality products (in the case of oil, tuna, or butter), and their consumption of nonlocal products, such as rice, bread, fish, and manufactured edibles, has risen. Local families are directing their temporarily higher income to increase their general consumption,[3] savings, the acquisition of assets (such as vehicles, land or properties), and more and better formal education for their children.[4] What the local families are not doing, however, is investing in farming production in the area. In a situation of high expectations for being resettled, local families are directing much of their efforts and hopes outside farming activities.

Even many of the landless farmers from surrounding localities prefer to press the mining company to be included in the project's area of influence; and, consequently, they benefit from the local procurement programs instead of renting fallow land. Indeed, in a context of decades of decline of the traditional farming sector, farmers would avoid long-term investments in agriculture. Analyzing the data of the national agrarian census of 2012, Pintado (2014) finds that the farming sector is losing its share of weight in the country's economy. To be sure, farming incomes are by far the lowest for farming and nonfarming households[5] and, as is to be expected, members of farming households are increasingly working in nonfarming activities. Indeed, the increase of nonfarming activities of rural households, together with illegal economy, explains much of the current reduction of rural poverty in the country, which is falling even faster than urban poverty. In La Granja's case, changes in the productive system—which involve a displacement from farming activities to nonfarming activities fostered by mining development—have significantly increased the local income, although this is temporary.[6] However, the displacement of labor outside of agriculture, the lack of technological inputs to increase productivity, and the increase of local consumption have led to a sharp increase in local prices. This situation especially disadvantages vulnerable families and those who do not benefit from mining-related opportunities.

5.2 Nonfarming activities

In a study conducted in the mid-1970s in the province of Cajamarca, Deere and De Janvry (cited in Velazco 1998) found that almost 55% of rural families' total income was derived from handicrafts, wage employment, trade, and remittances. Indeed, nonfarming activities in the region had been important before the analyzed changes. However, the attempts to develop the mining project have created large opportunities outside farming occupations in two principal ways: (i) direct employment in the project and (ii) creation of small businesses to provide services to the mining project (for instance, transportation, engineering and construction, or the lease of heavy machinery) or goods and services to the local families and newcomers (lodging, grocery stores, sale of oil and gas, restaurants and canteens, drugstores, billiards, among others). While the first option is linked to local procurement programs implemented by the mining company, the last is in response to rapid population growth in the area and the average increased purchasing power of the families. In any case, the families draw upon mixed strategies to benefit from the diverse opportunities that have opened up. Therefore, it is not surprising that in all of the seven examined families living in La Granja, at least one family member is directly employed, temporarily or permanently, by RTLG or by one of the many contracting firms. In addition, in three of the four interviewed migrant families, some of their members have returned to the area and been incorporated into the temporary employment program.

The labor history of Rafael, the male informant of the Resistant family 1, is a good indicator of the strategies surrounding direct employment. During the time of Cambior, he worked as an assistant in the material depot. When Cambior left the project, he returned to agriculture. However, when BHP arrived in La Granja, Rafael assisted in the exploration works. Later, with the gained experienced, he began to work as a master builder to construct houses for the returnee families. In 2006, when Rio Tinto took over the project, Rafael began to work as an assistant in the crusher unit. In 2009, through the *ronda campesina*, he was employed for eight months in the installation of the sewer system, a project implemented by the Social Fund La Granja. From 2010 to 2013, with an interruption in 2011 due to the financial crisis, he worked as a drilling assistant within the temporal employment program. This totaled two years and eight months of employment. After his work with RTLG finished, he returned to building houses, although he now hires personnel. Despite a rise in the demand for housing, he prefers working at the mining project because it provides a higher and fixed salary while in the construction business he has to share the benefits. In addition, Rafael and his wife opened a restaurant in 2003 in their first house. They ran the restaurant for one year, after which they opened it only seasonally during the local festivities until 2011. The business was an initiative of his wife as an additional income for the family. However, it demanded much work and she was alone. Nowadays, she lives in Chiclayo and is an independent worker in informal activities. With this history of long-standing and relatively successful engagement in nonfarming activities, it is not surprising that Rafael does not wish to return to work in agriculture: "the worst think that could happen to me is that the project would leave La Granja and I must return to the farming fields".

Certainly, the access to direct employment around the project activities is the most significant concern for the local population. The most important demand in the strikes of 2008 and 2011 in Querocoto, organized by the directives of some of the *ronda campesina*'s zones, was the provision of more employment opportunities for the entire population. In 2013, there was another strike against the company, which blocked the entrance road to the project in La Granja village. This time, a significant portion of the local population asked the company for higher salaries and better working conditions with Sodexo, a contracting firm in charge of providing catering to the mining camp. Indeed, there are two main issues regarding local employment: who is eligible and under which conditions.

The first issue has, in turn, an external and an internal facet. At the external face, it opposes localities that are inside the "core area"—which has been defined following environmental principles as the "direct impact area"—with those outside.[7] Unsurprisingly, localities outside the "core area" claim that they experience substantial negative consequences and pressure the company to be included in the local programs. These localities were the main supporters of the strikes in 2008 and 2009 in Querocoto. To manage this pressure, the company has used the Social Fund as a tool to provide infrastructure, development programs, and local employment to the outside localities. In addition, families at the "core area" would attempt to monopolize all the available employment opportunities at the project. For instance, they complain that outsiders are being hired despite the fact that they themselves are not skilled workers. Any job that is taken by an outsider represents a lost opportunity for a local person. Within the company, there is tension between the company's employability standards and local expectations of including the greatest possible percentage of the population. For instance, while the company has set a minimum and maximum age of 18 and 55 years, respectively, much of the local working force lies outside those age ranges. Likewise, for legal purposes, the company demands that each worker holds a national identification card. In a rural area with little presence of the state, undocumented people must go to the district to register themselves; this is especially the case for women, who are less likely to hold identification. Other requirements, such as the completion of secondary studies, or the obtaining of a professional driver's license, in the case of truck drivers, have provided incentives for people to finish their studies in nonformal programs for adults or enroll in training courses. Perhaps the most contentious requirement is health evaluation. Many people consider it unfair and discriminatory. Certainly, in poor, rural areas lacking effective public health services and existing under difficult working conditions, health issues are the most frequent reasons to exclude people from the employment programs; from relatively minor problems—such as dental care or short-sightedness—to more chronic conditions including tuberculosis, malnutrition, or spinal deformity, which is common among farmers owing to prolonged use of plows.

Regarding the conditions of employment, the contractor and the duration of the engagement are the two main considerations. The temporary local program set by RTLG provides job opportunities with the company itself and, mainly, the diverse subcontracting firms that provide services to the project. The local population, as much as possible, prefers to work directly for RTLG because they consider that the mining company provides a better working environment. The 2013 strike in La Granja was specifically targeted at one of the project's contracting firms. In addition, jobs at the mining project could be temporary or more permanent ones. Temporary jobs, which are part of the rotatory system, are offered to nonskilled workers and are managed through the *ronda campesina*.[8] More permanent jobs are directly negotiated with the company and require higher qualifications.

For many of the local families, permanent and inheritable employment is perceived as a right for giving up their lands. They expect that mining benefits (namely, employment) will last throughout their lifetime and would be inherited. In other words, the families expect to exchange an asset, which lasts for life (land) for another asset (permanent employment) for life. Despite low productivity and poor yields, land access and land ownership have become key assets to negotiate. In a sort of parallel, whereas for junior companies, the value of undersurface rights does not reside on the project development itself but in their ability to sell a dream of future vast yields (Tsing 2000), for the local families the value of the land lies not in its farming use but in its exchange value to negotiate benefits.

The case of Jorge, the male informant of the Returnee family 1, illustrates many of the complexities surrounding the creation of businesses around the mining project. For most of his life, Jorge has been a merchant. Before the arrival of Cambior, he farmed in the fields and traded cattle. He bought local cattle and then sold them at Chiclayo's slaughterhouses each week using his father's truck. When the exploration works began in the area, he worked as a truck driver for the company from 1994 to 1996, the year the family sold their land to Cambior and moved to Chiclayo. In the city, he continued with the business, feeding cattle to sell in Lima. In 2002, the family moved back to La Granja. In addition to cattle trading, he began to buy and sell granadilla and coffee. With a well-established business, he bought his own truck to avoid freight costs. When Rio Tinto arrived in the area in 2006, he bought a small truck from his father and, as part of a local business called Greenkart, he transported solid waste from the mining camp. From 2008 to 2010, when RTLG reduced its operations due to the financial crisis, Jorge transported construction materials in Cutervo. In 2010, RTLG reassumed operations and he decided to buy a water tank truck to provide services to the mining company. In addition, he formed his own business to directly engage with RTLG. In 2011, Jorge bought a light truck to be used as an escort in the daily convoys from the camp to Chiclayo. In 2012, he acquired a heavy truck to transport material from the mining camp to Lima. Nowadays, he manages the business and hires three drivers and one administrative assistant. Through his business, he provides services to the mining company, transports products from La Granja to Chiclayo (and vice versa, charging freight), and leases trucks to other people. Via this strategy, he obtains a regular monthly payment plus extra revenue from the leases and freights. Jorge claims that one of the disadvantages of owning a formal business is the payment of taxes, the accountable audits, and fines from the state tax agency. Indeed, as in the case of the local employment program examined above, the working scheme implemented by the mining company—which is based on a technocratic view of social, health, and environmental standards (Himley 2014)—frequently collides with local practices and expectations. One of these practices is the resistance of paying taxes to the state.

Jorge provides these services to the mining company through one of two local transport associations. The association, which split from a previous one, comprises 20 transporters from La Granja and Querocoto. With its headquarters in Chiclayo, the association was created in 2010 to ensure that larger transport companies from Cajamarca city and Lima would not provide services to the RTLG.[9]

These associations mediate between individual providers and the mining company and seek to prevent nonlocals entering into the contracting system. For its local supplier program, RTLG requires contractors to obtain a one-year warranty of functioning although three months is the maximum period of service. In addition, the contractors must employ local people and provide them with the legal social benefits. With the purpose to be part of the local suppliers, some families have sold their houses in the coast to acquire machinery and to implement the business according to the company's rules. However, many of them have not been called to provide services. In a context where many local families do not have enough capital to fulfill these requirements, some outsiders provide the capital and machinery under the name of a local person in exchange for a monthly payment. This is one of the forms to evade the monopolistic arrangement of the local supplier program. Otherwise, RTLG forbids that a person employed in the mining project could be a local supplier. In a similar strategy as the one used by nonlocal firms, some local people open their businesses using the name of others so they can continue working in the local employment program. These accounts illustrate the interaction between social forces and family strategies, which in their aim to maximize their benefits are changing the productive landscape of La Granja.

Agrarian economists investigating income strategies of Andean rural households find that when the farming land is too small—and other agrarian assets such as pastures, forest, or water are scarce—and households cannot generate enough income to cover their expenses, they employ two main strategies (Kervyn 1996). One is the use of seasonal migration seeking wage employment in the city or more productive agriculture areas (for instance, areas of coca or coffee cultivation). The other is a greater involvement in nonfarming activities within the agrarian unit. It must be noted that diversification, with different levels of intensity of economic activities and income sources, is a long-standing subsistence strategy implemented by rural Andean families. As a means of managing ecological niches to control ecological variation, rural families use economic diversification to reduce the strong externalities and risks that agriculture faces. With the increased development of rural markets and the more fluid interaction between rural and urban economies that the country has experienced over the last two decades, the importance of nonfarming activities—both in terms of income and allocation of labor force—has increased (Escobal 2001).

La Granja's case does not contradict previous findings. They provide evidence and a nuanced insight into the complex and flexible arrangements made by local families in the context of mining development. For instance, the case of the male informant's father of the Resistant family 2 exemplifies a situation where the control of significant agrarian assets (he owned 60 hectares of land in a district where around 70% of farming producers only possesses between half to five acres) provided the conditions for a further investment in farming activities. In most cases, however, local families access very limited agrarian assets and are forced to implement migratory and income diversification strategies. For some local families, mining development in the area has created opportunities for greater income diversification and monetary benefits through the allocation of time and resources in nonfarming activities. If the mining cycle declines, many families will return to farming activities as they have in the past. In a fluid movement between farming and nonfarming activities, a strategy of the local families is to maximize employment opportunities in the context of subsistence farming economies.

If we agree with David Harvey's statement that "capitalism creates a physical landscape [...] in its own image" (1985, p. xvii), then the double residence pattern that many families from La Granja have developed could be understood as being the result of mirroring the particular productive configuration of the region. The residence in the city provides a beachhead from which to take advantage of urban and improved employment opportunities; however, its location on the city's outskirts and the housing and neighborhood, with their undesirable features, indicate the poor quality of the accessed employment. The residence in the country secures a safe base to which they may return in cycles of economic slowdown. In addition, their status as "locals" allows the families to access preferred employment and business opportunities associated with mining development. The recent construction of multilevel houses of brick, concrete, and glass reflects the extraordinary influx of capital into rural areas as a product of mining development. Moreover, not only does the physical landscape replicate the circulation of capital, but people also migrate as a result of this movement.

5.3 Transformation of the local power landscape

Unlike the case of land access, which depends more on individual family decisions, the way to access local employment at the mining project is often mediated by the collective space of the *ronda campesina*. The *ronda campesina*s emerged in Cajamarca in the 1970s as a mechanism to deal with banditry and cattle rustling. In a situation where the state is absent and there is a lack of strong social organizations, the *rondas* have become significant actors in the local and regional political scenario and mediate between families and external actors, such as mining companies or state representatives. The *ronda*s have formed a dense structure that goes from bases at village level to regional committees. In Cajamarca, the *rondas* have split into two different entities: the "central of rondas" and the "federation of rondas". And, although they are autonomous from political parties, the influence of Red Fatherland—a radical leftist party with a stronghold found among public schoolteacher unions—over them is notorious. In the region, the major part of the *ronda* bases tend to oppose mining development, especially in the Bambamarca province around Newmont's Conga project. The villages involved with La Granja project are grouped within zones 2 and 3 of the *ronda* federation, and their bases play a more tolerant role in mining development than their counterparts in other provinces.

When Cambior arrived in the area, the company hired local personnel through the *ronda*. Nonetheless, because of the low and inconsistent company social standards and the weak organization of the *ronda*, company managers frequently bypassed the mediating mechanism. The company representatives preferred to develop a personal and patron–client relation with the population. As a local villager noted: "During Cambior's time, the work was for short periods of one week or fifteen days, but if you got on well with the engineers then you would stay longer". After purchasing the project, BHP Billiton also engaged with the *ronda campesina* as the main collective representative body. The general manager of the project explains: "The experience in Antamina mine suggested to me that the *ronda campesina* had a similar role to the peasant community and it articulated the social and political relations in the villages".

Rio Tinto's managers opted for a similar approach to engage with the local *ronda campesina* bases from the early stages of the project.[10] The situation of the local *rondas* was not of institutional weakness at the time of RTLG's arrival. However, the reactivation of the project impelled the *rondas* to assume new and diverse tasks, which resulted in their revitalization. They shifted from acting as patrol organizations to playing a central role in the agreement negotiations concerning employment, local supply, development projects, and resource allocation. In this sense, they have become a parallel power to the local government. The increasing legal and organizational flexibility that the *rondas* enjoy, compared to that of the local governments, has allowed them to grow and transform into a major organization in the channeling of economic and political resources for the development of their areas of influence.[11]

Consequently, in La Granja, the *ronda* organization is the main filter for the recruitment of local people in the temporary employment program. RTLG presents its job requirements to the *ronda* and the steering committee select, from its register, suitable applicants who meet the requirements to fill the available positions. It is not surprising, then, that the number of members of the *ronda* bases in zone 2—where the localities of the "core area" are located—has grown by 242%, from 190 in 2001 to 650 in 2013. By contrast, the number of *ronderos* in zone 3—out of the eventual resettlement area, but still in the project's direct area of influence—has grown only 36%, from 377 to 513 in the same period.

The company's strategy of engaging with the *ronda* has certainly affected its functioning, and some researchers (Damonte 2012) argue that the presence of multinational mining corporations in the rural scenario has the potential to deeply alter existing local social and political organizations:

> ...the presence of large-scale extractive companies produces three local phenomena regarding the forms of local representation [...]. Firstly, it causes that the communities or the *rondas campesinas*, generally marginalized, acquire higher visibility and prominence in the district and province political milieu. In the second place, it exacerbates the competence and fragmentation among the communities or *rondas* and within them; the position with regards to the mining operation is the cause of conflict and sometimes even fragmentation of the local rural organizations. Finally, the introduction of corporate regimes transforms the nature of the organizations and the forms of local representation, because they became inserted into the logic of the extractive project.
>
> (Author's translation, italics added, p. 112)

Damonte's arguments highlight the enormous capacity of large-scale projects to alter the social and political dynamics of small rural localities. However, it is worth noting that mining investments could also strengthen local institutions that have had a poor performance and minimal relevance before their presence. Before the development of mining activities in the area, the *rondas* were almost inactive in some villages. In addition, there are two competing structures, each with different interests, capacities, and functioning. Certainly, as a result of the lack of presence of state and central government institutions, the extractive project needs to engage with local institutions; and, by this process, it shapes their function (for instance, from patrol to recruitment agency) and internal dynamics (for instance, increasing fragmentation and conflict). In the language of corporate social responsibility, the project has used and reinforced the existing social capital. However, as I have argued in the case of indigenous organizations in the context of land-titling policies and programs promoted by the World Bank in Latin American countries in the 1900s (Castillo 2002), local forms of organization could potentially benefit from external forces and institutions. As an unintended consequence, these organizations could increase their own demands and interests and become accredited interlocutor face-to-face state agencies and extractive companies. In Peru, for instance, the passing of the Prior Consultation Law in 2012 by Congress has reinforced a process that requires the identification of legitimate social and political indigenous representations. In brief, the performance of the *rondas campesinas*—as for any other local social or political organization—is less the automatic result of the external intervention of a mining project than the complex intertwining with local dynamics of power.

Gender relations are another power dimension that is being transformed by ongoing mining development. In La Granja's agrarian society before its current transformation, women were excluded from land ownership and lived in a patriarchal system—first under the authority of their fathers and then of their partners. From childhood, they were involved in labor activities, both on the farm (specially to graze the flock) and at the house (to undertake domestic duties). These productive and reproductive tasks, however, were not socially recognized, and in the main, they were unpaid. Through different mechanisms— not necessarily designed with a gender-sensitive approach—mining development is altering local gender arrangements.

First, because of the economic bonanza, men are more prone to leave farming activities in search of higher salaries in other productive sectors. Factors including higher education and training levels or the migratory experience of men partially account for their increasing participation in nonfarming activities. Power relations, nevertheless, better explain why men tend to reap the economic benefits and wage employment opportunities. Similar to what has been found in other parts of the world (Farrell et al. 2004), women have received fewer economic benefits from the last three mining companies that have operated in the area as the majority of them are not legal landowners; and, thus, the firms have tended to directly negotiate with men. In addition, the collective interlocutor—the *ronda campesina*—has for a long time been an exclusively male-dominated organization. In many places, this situation has led to an increase in women's economic dependence on their partners. In a research about the effects of tax revenue distribution system in the eastern province of La Convención in Cusco, where the Camisea gas project is located, Viale and Monge (2012) found that, with the declared purpose to avoid a labor shortage on the regional coffee plantations, the local government excluded women from the temporary employment program that the municipality had implemented. Nevertheless, the "feminization of the agriculture" (Remy 2014) is challenging some of the power structures of gender relations. Although women's work in farming activities is viewed as complementing the family economy, women are increasingly appropriating part of the farming gains and raising their voices for collective representation in the locality. Partly because of the policies and practices of antidiscrimination implemented by RTLG,[12] women are included in the temporary employment program, and they have formed a women's *ronda campesina* with the purpose of managing their recruitment process.

Consequently, and in the second mechanism, women are gradually accessing wage employment opportunities presented by the mining project. Women, who are mainly employed in cleaning and cooking tasks, comprise around a quarter of the local workers at the RTLG project. In addition, a significant part of this percentage is represented by single mothers or single women. This situation is not necessarily owing to a special focus on vulnerable individuals through the employment approach, but because childcare and domestic tasks severely reduce women's possibilities for accessing and securing a wage job. As a local woman explains: "Work in the mine is for single women, when women have family, they neglect their duties at home and the care of their children. I had to resign to take care of my family". These conditions show the reproduction of expected labor roles for women and the hidden barriers—which tend to be naturalized—they must face for accessing a wage employment. Certainly, the unequal distribution of work at the domestic level has not been challenged. In some cases, the family hires a person to assist with domestic tasks, generally a teenager or an old woman from neighboring localities or areas, considered more rural. The latter situation would tend to perpetuate and exacerbate relations of poverty, age, and ethnicity among women. The other option is that women working outside the household bear a labor overload or transfer part of these duties to their daughters.

In addition, women are increasingly participating in local business. Most of these businesses (including restaurants, grocery stores, canteens, and laundries) are regarded as an extension of domestic duties, which implies restricted mobility outside the village. Despite these limitations, women are attaining better education levels and greater economic, physical, and political autonomy. Without a doubt, local ongoing progress in terms of women's autonomy is met with great resistance and results in an increase in gender tensions. Perhaps the most vivid examples of these tensions are the accusations of infidelity, divorce, and family separation by men and women alike, leveled against employed women. For example, a mature local woman declares, with indignation:

> Women that have a wage employment or a business earn more money. This situation has meant that women are more liberal now, so they are unfaithful to their partners and then get divorced. Because there are so many newcomers, women regard their husbands of little worth.

The explicit link between women's economic autonomy and their sexual liberty and free circulation among men reveals the challenges that women must confront in a transforming society. From the age of 15, the female informant of the Migrant family 3 had to leave her hometown because of the difficult economic conditions experienced by her family. She went to Chiclayo where she got a job and made her life as an independent woman with her own friends. Some years later, she became engaged to a man from La Granja and gave birth to a daughter. The man has occasional jobs in the city, drinks heavily, does not help her with childcare, and tries to control her life. For these reasons, she wants to return to work and separate from him. Because of its greater job opportunities, local businesses, and better transport facilities, she prefers to live in present-day La Granja, explaining that, "only old people like how La Granja was before".

Notes

1 The American mining company, Newmont, together with the Peruvian group, Buenaventura, controls Yanacocha, one of the largest gold operations in the world. Since its arrival in the region in the early 1990s, Newmont has faced strong opposition to its controversial expansion plans in a series of episodes—Expansión Oeste, Cerro Quillish, and, more recently, Conga.
2 Though RTLG opened employment opportunities for women, and the *ronda de mujeres* negotiates with the company a certain number of temporary employment positions, the percentage of men working in the project was considerably higher.
3 Of course, the levels and the patterns of consumption are quite different between and within families. There are also significant gender and age differences in these patterns. While adult women tend to invest their money in clothing and food for the family, young men will prioritize prestigious goods, such as branded clothes, mobile phones, motorbikes, and beer, which is replacing the traditional and cheap sugar cane liquor.

4 The number of children that complete their primary and secondary level has substantially increased in the area, and many of them follow their secondary studies in Chiclayo, where there are higher qualitative levels. In addition, there is a significant group of youth that has followed technical and university studies, the majority of whom study in private institutions in Chiclayo. A group of these young professionals has formed GRAPAMI, an NGO that has implemented leadership and education programs among teenagers in the area.

5 Thus, for instance, the monthly national average income in farming activities for farming households was US$102 in 2013. For trade, transport and communications, construction, services, and mining activities, the average incomes are: US$ 160, US$ 255, US$ 293, US$ 334, and US$ 359, respectively (Pintado 2014, p. 4).

6 In La Granja, the average monthly income increased by 120%, from around US$ 265 in 2010, a year before RTLG reassumed the temporary employment program, to US$ 610 in 2013 (Consultant informant 2).

7 Ten localities compose the "core area". This definition goes back to Cambior's management of the project and has been later validated by the state. First, with the Social Fund agreement signed with Rio Tinto and later, with the preparation of the Environmental Impact Assessment of the project.

8 At the beginning of the program, the positions offered were for a one-month period. Later, through a negotiation with the *rondas campesinas*, this was extended to a period of three months, so that the worker would be able to access health insurance.

9 The other association comprises 36 members. The high number of members of the original association slowed returns for each supplier. Together with conflicts over control of the board, this situation prompted the division of the association.

10 This decision responded to the need for establishing an interlocutor with whom to negotiate, but also to limit the social commitment as much as possible to answer the criteria of environmental direct affected area. Therefore, the project did not initially engage with the municipality of the Querocoto district. Nevertheless, a major strike and road blockage in 2008 forced the company to include the local government in its social programs.

11 Local governments are constrained by legal and administrative rules and, in mining areas, have been overwhelmed by exponentially higher financial resources without additional and more efficient human resources. On the contrary, organizations like the *rondas campesinas* operate at the margins of legality; they do not pay taxes, are not audited, and the projects they manage are free of the state procurement regulations.

12 Rio Tinto is one of the few extractive companies in the world that explicitly considers the gender effects of its operations and which has developed policies and protocols (Rio Tinto 2009).

6   Mobility

Since early sociological works in the late 19th century, migration has been regarded as being at the core of the transformation of rural societies. To be more precise, reflecting upon the West European and North American experiences, social scientists regarded migration as a unidirectional movement of people leaving rural areas to settle down in the newly industrialized cities. Nevertheless, the postcolonial experiences in Africa and Southeast Asia and the large urbanization processes, which unfolded in Latin America after World War II, questioned the necessary link between rural–urban migration and industrialization. The growth of large shantytowns and the persistence of high rates of unemployment and subemployment (Roberts 1978) are the most notorious examples of processes of urbanization without industrialization. In other words, many societies become urbanized without transforming their economic structure, which continues to be heavily dependent on the primary sector, especially agriculture. Interestingly, social studies concerning mining areas in Africa and Latin America were more sensitive to the complex and fluid directions of rural–urban migratory flows. They analyzed, for instance, seasonal migration, complementary household economies between farming and mining activities, and the gender imbalances and consequences of male movement to mining towns (Godoy 1995; Ferguson 1999; Bridge 2004; Damonte & Castillo 2011).

In Peru, social studies on rural–urban migration date back over more than 40 years and have formed a well-established research area. From historical standpoints, these studies have explored the formation of regional circuits (Bonilla 1974; Assadourian 1982; Contreras 1988; Contreras 1995), the establishment of seasonal migration patterns in regional economies (Long & Roberts 1984; DeWind 1987), as well as the dislocation of economic spaces (Manrique 1987; Helfgott 2013). From an anthropological perspective, the first research studies coincided with the first waves of massive migrations of farmers to the cities in the coast, particularly Lima, in the late 1950s. They primarily focused on adaptive strategies of "peasants in the city" and cultural reproduction (Altamirano 1977; Wallace 1984). Influenced by the work of Oscar Lewis, these studies were followed by an emphasis on the "culture of poverty" and the examination of the deleterious effects of displacement, in terms of social anomia and cultural loss (Sandoval 2000; Vega-Centeno 2006). In the 1970s, after two decades of *campesino* immigrant struggles for appropriating and building their own space in the hostile cities, the research attention shifted to the scrutiny of the reproduction and activation of social networks in the urban spaces (Rodríguez, Riofrío & Welsh, 1973; Altamirano 1977) and the effects of migratory flows, both in the localities of departure and the receiving cities (Fuenzalida et al. 1982; Alber 1999). With the return to the democracy in the early 1980s, various studies highlighted the emergence of political consciousness and citizenship building among former immigrants (Degregori, Lynch & Blondet 1986) and their entrepreneurial strategies (Golte & Adams 1987). At the turn of the century, researchers have broadened their scope to include transnational migration of Peruvians in the United States, Europe, Argentina, and Chile (Altamirano 2000; Berg & Paerregaard 2005), where they explore issues of cultural reproduction, activation of social networks, or changes in the localities of origin. Considering a third generation of people born in Lima of immigrant parents, current researchers concentrate their efforts on the identification and analysis of segmented cultural patterns of consumption among diverse city dwellers (Uccelli & García Llorens 2016), the formation of popular culture styles (Huerta-Mercado 2006), or the persistence and reproduction of segregation using racial and sociocultural codes (Nugent 2012).

In addition, there has been the development of a series of works under the loose framework of the "new rurality" (Giarraca 2002; de Grammont 2004; Cetraro, Castro & Chávez 2007). These studies argue that rural economies are no longer exclusively or mainly associated with farming production. The separation of the agriculture and rural settings is mainly due to the increasing importance of services (including tourism) and the formation of an interconnected society. Vast improvements in communication networks—which allow a more fluid flow of information, goods, and people—have made it possible to narrow the distance between rural settlements and urban areas. Thus, it makes little sense to think of rural spaces as autonomous physical, social, productive, and conceptual entities separated from the cities. They are part of larger spaces that connect metropolises, intermediate cities, and local places in networks of different scales (Allen, Massey & Cochrane 1998; Hurtado 2000).

A deeper understanding of the historical trends of seasonal and permanent migratory dynamics enhances our ability to comprehend the family strategies in the context of extractive development. What follows is an analysis of the mobility processes through which the *Granjinos* have created a dense spatial network over time.

6.1 Migratory experiences

The examination of La Granja's case shows that the families in the village have a long-standing experience of emigration and immigration processes. Members of diverse families have moved outside the area in different moments for different reasons, and many of them have also returned to the locality. In the contemporary local history, it is possible to identify three major migratory cycles.

The first cycle goes back to the years prior to the development of mining activities in the early 1990s. Because of the pervasive lack of paid jobs, the low returns of subsistence farming, and the difficulty for continuing subdividing their already small farming units, many families opted to send some of their members from the locality, especially young people. Men tended to migrate to coffee and coca plantations in the lowlands, or to the coast to work in agriculture or to seek temporary employment in the construction and service sectors. Women mostly migrated to the lowlands to work in the farming sector and to coastal cities, where they became domestic workers. This has been the case of the Migrant family 4. Holding very few properties, the family sold their scant belongings and moved from Paraguay to Ojo de Toro, a rural area on the coast. Indeed, as this family did, thousands more from the Cajamarca Region migrated to the coastal regions of Lambayeque and La Libertad, trying to escape from poverty. According to the national census taken in 2007, Cajamarca ranked as the fourth most populated region in the country with 1,387,809 inhabitants, of which 67.3% lived in rural localities; this was slightly behind Cusco as the region with the highest portion of rural population (Instituto Nacional de Estadística e Informática 2007). Poverty levels in the region remain among the highest in the country, ranking seventh in 26 regions, with 32% of Cajamarca's population living in poverty (Instituto Nacional de Estadística e Informática 2012b). Given this situation, it is little wonder that Cajamarca ranks as the region with the second highest negative migration flows in the country, just behind Huancavelica, the poorest region in Peru. During the period between 2002 and 2007, the region exhibited a negative net migratory balance of 86,804 people, which represented a net migration rate of −13.6 (Instituto Nacional de Estadística e Informática 2007). It is worth nothing that this trend is almost identical to that of the period between 1988 and 1993 before the mining boom in the region (Instituto Nacional de Estadística e Informática 1993). In addition, the average annual population growth rate of Cajamarca for the years between 1993 and 2007 is only 0.7%, the second lowest Peru (Instituto Nacional de Estadística e Informática 2007). In brief, historically Cajamarca is a relatively well-populated region, heavily rural, with high rates of poverty and significant levels of emigration. Current mining development in Cajamarca, which began with Newmont's gold project, Yanacocha, in the early 1990s, has not reversed these trends but has shaped specific local processes.

A second critical moment in La Granja's migratory history was the massive displacement, a product of the implementation of Cambior's land purchasing plan of the mid-1990s. Unlike previous migratory movements, on this occasion, there was a lack of free will. Local families had very little option but to sell their land and settle elsewhere. The move was not part of a well-planned family strategy but was a compulsory event forced by many legal and intimidating mechanisms implemented by the mining company and the state, under the authoritarian Fujimori regime.

Like many others, the case of the Returnee family 2 illustrates many of the stressful components that accompanied the displacement experience. After selling their land, the family moved to Batán Grande, a semiarid area in the Lambayeque Region, three hours from the city of Chiclayo. Prior to moving, the family made arrangements to build a house in the new area. However, they were cheated and when they arrived, there was nothing in the terrain. As Blanca, the female family head, narrates: "…it was pretty sad, there wasn't any house, we had to build a provisional hut". The heavy rains of the El Niño phenomenon of 1997 severely affected the area and the family lost their crops and their precarious house. Blanca recalls that "it was a nightmare, thousands of mosquitos, we had never experienced something similar in La Granja". The family received social support from a religious organization and was temporarily hosted in the local school. In the aftermath of the rains, a cholera epidemic spread throughout the region, infecting two of the family's children and killing Blanca's mother. In addition, the farming system on the coast is different from the one practiced in Andean regions: it needs irrigation mechanisms, uses another type of plow or tractor, requires pesticides, fertilizers, hybrid seeds and other inputs, its crops exhibit different vegetative cycles, are market-oriented (i.e., corn for chicken factories), and are not necessarily suitable for direct consumption. Until the family learned and adapted to the new productive system, they faced hunger. Moreover, the family felt the lack of social bonds and the hostility of a different culture, which they regarded as being more individualistic and competitive: "We had cattle, but the people were envious, and they spoiled the udders of the cows. People are selfish, they could eat in front of you and offer you nothing".

An engineer advised the family of the great risk of the La Leche River overflowing, and they decided to sell their land. After nine years on the coast, the family returned to La Granja by their own means; they did not receive the support of the BHP returning plan. The family has restarted its life and many of the family members have reunited again in the area. However, the displacement was a disruptive event that dislocated the family. The male informant's mother-in-law and father died in Batán Grande, two of his brothers remained there, while some of his sisters moved to the lowlands in the eastern part of the country.

The third migratory wave has been prompted by a new cycle of mining development in the area with the arrival of Rio Tinto in 2006. The implementation of temporary employment and procurement programs, the creation of a developmental social fund, and the opening of small-business opportunities have created great incentives for the arrival of a significant number of immigrants to the area. Certainly, from only 7 or 12 houses that remained in La Granja after the displacement caused by Cambior's land acquisition plan, there are now between 140 and 180 houses. Schoolteachers, health workers, consultants, merchants, and landless farmers from diverse rural localities in the region and unemployed people from urban places have migrated, alone or with their families, to the area seeking higher incomes. Nonetheless, many newcomers are not completely foreign to the locals. On the contrary, many of them are relatives of the local families.

Amid the changes prompted by mining development, families from La Granja use kinship networks to promote but also regulate migratory flows. In this way, when job opportunities decline in the locality, many of the family members migrate to urban centers on the coast or move to more productive farming areas in the eastern lowlands. Nevertheless, some members remain in the locality with the purpose of securing land and properties. These latter members tend to be the elders. When opportunities for jobs and better salaries arise in the locality, members of the extended family networks move back temporarily. The case of the Resistant family 2 provides some examples of this strategy of spatial and temporal mobility. The male informant and his ten siblings are originally from La Granja. Of those 11, one (the informant himself) has remained in the village, two of his sisters married men from the nearby village of La Pampa and settled there, and eight have migrated to Chiclayo or other coastal cities. In addition, the family has six children. The youngest two live in Chiclayo, where they undertake tertiary studies. The other four also lived in Chiclayo, three undertook technical studies there, and the other enlisted in the police department. However, all of them have since returned to La Granja and are employed at the RTLG project. The female informant travels every 15 days to visit their youngest children in the city and the male informant makes regular trips to Chiclayo because of his business activities.

These complex movements have built a fluid and dense network that connects individuals and families in the broader space. As a consequence, Cajamarca's families have created concentrated pockets of fellow immigrants in the bordering regions. In Ojo de Toro, for instance, the families from Cajamarca have by far outnumbered the native population and they have recreated their own cultural (for example, religious festivals), social (godfathership, for instance), and political (namely, the *ronda campesina*) practices. The map shown in Figure 6.1 seeks to illustrate the spread of the migratory grid developed by the examined families from La Granja over time. The grid runs from localities in the vicinity, such as Pariamarca, Querocotillo, Pagaibamba, Mitobamba, and Querocoto, to the regional scale, from the city of Chiclayo on the coast to Tarapoto in the eastern lowlands; and it even extends to international places, with Quito, Cuenca, and Loja in Ecuador, and Buenos Aires in Argentina as the main destinations.

Thanks to the employment of this strategy of high spatial mobility, families are able to maximize economic benefits, retain strategic decisions inside them, and nurture personal bonds of trust. In addition, families use marital alliances and symbolic kinship (godfathership) as strategies to incorporate external members into their networks. It must be noted that in order to participate in the temporary employment program established by RTLG, which is one of the main reasons for migration to the area, people must belong to the *ronda campesina*. The *ronda* assembly is the body that confers membership, and to become a *rondero*, a person must be part of one of the local families.

The migratory record of the Opportunistic family 1 highlights some of the complexities involved in the decision for a nonlocal family to migrate to the area. The couple is originally from Querocotillo, a district capital in the province of Cutervo, in the Cajamarca Region. Because

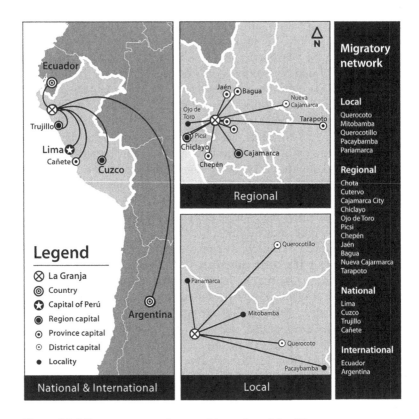

Figure 6.1 Migratory network map of interviewed families.
Source: Castillo (2015, p. 132).

of limited employment opportunities in the rural district, the female informant migrated to Chiclayo in 1965, when she was 23 years old, and became a domestic worker. The male informant followed a similar path and moved to Chiclayo, where he started to work as an informal merchant between different towns of Cutervo and Chota, in the highlands, and Chiclayo, on the coast. In 1969 the couple got married and that same year, their first daughter was born. In the early 1970s, the male informant included La Granja in his trading route and began to make regular trips to the village. By that time, the family had already moved back to Querocotillo, where the female informant inherited a small piece of farming land in the locality. From 1983 to 1991, the couple embarked on a small and seasonal business in La Granja, selling food in the streets during public and religious holidays. However, business was low and in 1992, the family decided to sell their properties in Querocotillo and move back to Chiclayo, where they bought a house. The same year, a son of the couple moved to La Granja and bought a house, which he sold to Cambior during the displacement process, and migrated to Chiclayo. In 2001, the second daughter married a *Granjino* and moved to the area. With these already established connections and the familiarity they gained with the area, the couple decided to buy an urban plot in La Granja and build a house. In 2003, they moved to the village.

According to the male informant:

> …we decided to move to La Granja because there were not many working opportunities in Chiclayo and crime was high. In addition, there were rumors of the arrival of a mining company in the area; so, we ventured to buy an urban plot, it was an opportunity to get jobs for our children.

Nowadays, five of the eight children live in La Granja; two live in their parents' house and three have their own places. The remaining three children live in Querocotillo, Chiclayo, and Lima, respectively, with their own families. In addition, the five children living in La Granja work in mining-related activities, four are directly employed at the RTLG project, and one has opened a transport business with his father, which provides services to the project. The couple has never been fully engaged in farming activities; they are mainly dedicated to trade in the bigger northern region of Peru. Otherwise, they have an important migration experience in Chiclayo.

6.2 Spatial dynamics

The experiences of these diverse families examined suggest that migratory practice is not an automatic and direct consequence of mining development in the region. From a family perspective and from a historical standpoint, migratory practices are better understood as an extended strategy of regional dispersion/connection, which seeks to enhance family economic advantages in the context of high risk and vulnerability—both in ecological (for instance, droughts) and socioeconomic terms (for example, a fall in the relative prices of agrarian products). In this sense, the economic and social rationale of migratory patterns mimics the strategy of vertical control of ecological niches that households and larger social units practice in the Andes (Hirsch 2017).

In addition, migratory experiences cannot be reduced to economic and livelihood strategies. They also involve the reshaping of collective and personal identities in interplay between what is regarded as rural and urban, traditional and modern. Furthermore, local expectations for the future are strategically elaborated in relation to the specific process that La Granja's families are experiencing with the last cycle of mining development. In this respect, in the period between 2006 and 2013, during the peaks of local employment and economic opportunities that the mining project development triggered, many members of local families have had the opportunity to reunite back in La Granja. Therefore, claims made by some local people that the development of the mine will destroy family unity would be better comprehended in the context of this period of immigration and family reunion in the area. Certainly, in a phenomenon not seen since times of the *hacienda* and in a significant larger scale, La Granja has turned from a place of expulsion to a place of reception, which brings new opportunities, fresh ideas, and diverse worldviews but which also entails resistance and new types of tensions. Indeed, the fast demographic transformation of the village has brought many young people and newcomers with different ideas, values, and lifestyles; and they are not necessarily part of the local kinship networks. These migratory flows and social changes create a sense of mistrust among some local villagers, a situation that is exacerbated by the uncertainty of the mining project and, therefore, the uncertainty that local families display about their lives. And an uncertain future tends to create anxiety. Like other mining regions, in La Granja the volatility of spaces of capital extraction is vividly manifest, both in terms of social bonds and physical materiality. Marshall Berman's famous use of Marx's phrase, "all that is solid melts into air" becomes dramatically real.

In this sense, the concept of diaspora might be useful for a better understanding of some of the features of people's mobility in the context of mining development. In its most general form, diaspora refers to ethnic populations living in places other than their historic homelands and with different collective practices, a situation that challenges the traditional boundaries of the nation-state (Lovell 2010). In La Granja's case, ethnicity and transnational migration are missing elements from the diaspora framework. Nevertheless, there are other features that shed light on some of the processes experienced by many *Granjinos*. For instance, for James Clifford, oppression and the trauma of departure is a defining feature of the concept of diaspora: "...without oppression to propel people to leave a place and to disperse there would be no diaspora" (cited in Kirshenblatt-Gimblett 1994, p. 342). Certainly, the displacement during Cambior's era was forced and traumatic for most families and it is perceived with a sense of injustice, especially considering the performance of a state enthusiastically supporting the company's actions.

Another component is the tension and interplay between "roots" and "routes" that the immigrants establish (Clifford 1994). This is to say, on the one hand, the tension between a sense of belonging and the need to travel, and, on the other hand, the continuous connections with a physical and imagined homeland. These provide the material for the development of nostalgia and an exercise of memory recreation, through diverse, fluid, and changing flows of people and ideas. This fluid movement created the possibility that "separate places become effectively a single community through the continuous circulation of people, money, goods and information" (Clifford 1994, p. 303). Certainly, the *Granjino* families that migrated to different places have reconstructed a social and imagined community despite the territorial discontinuity. Of course, the routes and connections are not without contradictions and resistances. For example, there are tensions between the reproduction of former social relations and cultural values and the desire for adopting the new social and cultural styles from the receptive society. This tension is often perceived and experienced as a clash between generations but also as a case of "divided loyalties" (Mitchell 1997, p. 534) between those who seek to maintain the traditions and those who pursue becoming modern. In the context of mining development in Peru, public demonstrations of "authenticity" and "tradition" become central elements in the political struggle for accessing benefits and compensations. Hence, individual immigrants will seek to show that they belong to the traditional culture and society with the purpose of being accepted as members of the recognized local organizations (namely, the *ronda campesina* and the *comunidad campesina*). Collective organizations will pursue proof that they are traditional and native organizations to be included within the official list of indigenous organizations and, therefore, to be able to claim the right of free prior consultation.

In a similar vein, experiences of displacement and resettlement provide a fertile field for the examination of the loss and remembering of social practices and cultural meanings, for instance, the recreation of social institutions, such as godfathership and the *ronda campesina* in different geographical settings. In addition, the notion of diaspora reveals the ambivalent space that displaced people occupy as cultural minority groups. In this sense, it offers a theoretical tool with which to explore the social and cultural frictions between host coastal populations and immigrants from the highlands.

Gender dimensions also play a substantial role in shaping spatial dynamics. Certainly, within this process of increased mobility, young women are experiencing levels of liberty for spatial and social mobility not seen before, although it is true that men still enjoy a higher spatial mobility than women do. Many of the analyzed narratives of spatial mobility emphasize the fact that while male partners often travel around different localities at local, regional, and national levels and for longer periods, female partners find it more difficult to leave the village. The trips that women make tend to be shorter in distance and duration. Of course, this pattern significantly limits women's migratory experience, in particular, and a broader social involvement, in general. The prevalent sexual division of labor—which makes women responsible for most of reproductive work—is the main barrier for women's mobility. Therefore, interviewed women declare that it is difficult for them to stay away from the village for many days because they must take care of the children or the elderly and fulfill domestic tasks, including care of the small livestock and the family gardens. Furthermore, men continue to dominate access to local public spaces. The examination of the local spatial routines indicates that for many women, their daily mobility is still confined to their homes, family gardens, areas for grazing the flock, or their small grocery shops and restaurants, which are perceived as an extension of their domestic duties and a physical appendix to their homes. On the contrary, the spatial mobility of men is broader and includes their homes, the mining camp, the farming fields, the roads and—more significantly—the street and the sports field, which is in the central part of the village. These routines not only reveal that women have more limited spatial mobility in their daily lives but also that the quality of those routines is more confined to social interaction among peers of the same age and gender.

Of course, women have their own spaces for socialization, such as volleyball games, church gatherings, and *ronda de mujeres* meetings. Moreover, women's rights and voice over the access to public space are becoming stronger. Nevertheless, those spaces are more limited and enjoy less prestige and power than the male spaces do. For instance, women play volleyball and use the village's sport field at marginal times, when men are working, and usually not during the weekends. In addition, soccer—which is played by men—is more prestigious than volleyball. For instance, during the saint festival of the Lord of the Miracles—the major festivity in La Granja—a soccer contest was organized, which gathered around 12 teams (some arriving from other Cajamarca's provinces and from Lambayeque), and offered around US$ 1,300 for the winner and US$ 650 for the second place. Women participated in the volleyball contest, which gathered no more than four teams from the village and neighbor localities and provided prizes of around US$ 100 and US$ 65, respectively.

Women also gather around church activities, mainly for praying and decorating the chapel with flowers, which are activities that men do not consider particularly important. However, a man presides over the Asociación Señor de los Milagros, a central organization because in addition to its religious activities, it still owns diverse plots in the township. Finally, the *ronda campesina*—composed exclusively of men—have far more members and power than the *ronda de mujeres*.[1] This is partly explained by the fact that the *ronda de mujeres* is a sort of artificial organization created by the initiative of RTLG with the purpose of organizing and selecting women for its local employment program. According to some informants, more men than women work in the RTLG local programs, at a rate of seven to three. The latter situation is not only caused by the company offering fewer jobs to women. It is mainly because women face more difficulties in managing a wage employment while still being responsible for their families' reproductive duties. Thus, it is no surprise that most of the outsiders seeking jobs in La Granja and trying to become members of the *ronda campesina* are men. Women, and especially single women, are more restricted in their physical and social mobility.

Within this context, many adult women tend to interact mostly with their children and other family members daily, which may foster a sense of isolation and reduce their chances of developing strong peer solidarity. Men, on the other hand, develop much of their interactions with other men at work and in recreational spaces. It tends to be around drinking alcohol, playing soccer, gambling on cockfighting, listening to music, or gathering in the streets that men generate strong gender solidarity. The street symbolizes the space for the formation and reproduction of the male brotherhood, and it is the place where men are able to accumulate substantial social capital and prestige (Fuller 2001). Furthermore, men's appropriation of the public space leads to the reproduction of a binary opposition between a feminine home and a masculine street/working space.

Despite these considerable and enduring limitations, women are currently experiencing higher spatial mobility than some decades ago. As a local woman narrates:

> In the old times, young women did not travel alone, but always in the company of someone. In the case of many young women, they moved out from the village to work as housekeepers in cities in the coast, and they went to the houses of close relatives. To travel

alone, they had to ask permission of their parents, because even though they were adults they were still living at their parent's home. In some cases, where women had started living together, they sought permission from their partners.

An elderly female interviewee, for instance, moved from Querecotillo to Chiclayo when she was 22 years of age to work as a housekeeper in the home of one of her cousins. Nowadays, she prefers to live in Chiclayo instead of La Granja. This is not only because she receives regular medical treatment in the city but also because "La Granja is boring; I have nothing to do here. When I am in Chiclayo I can stroll by the streets with my daughter and son, I can go to parks and do shopping. I get amused in the city".

Nowadays, women have more opportunities and fewer constraints in terms of moving out of the village and going to different localities in the region. Some of them have even developed a substantial migratory record, which includes cities outside the region—such as Lima or Cañete—and even other countries, Buenos Aires for instance, following the experience of thousands of Peruvian immigrants seeking better job opportunities.

In addition, today women face fewer risks and resistances when freely moving within and using the public space and streets of La Granja than in previous generations. As one local woman declares: "... nowadays, women freely walk in La Granja, the place is safer than it was before. There are less fights and drunk men". Indeed, these days it is not a surprise to see groups of girls hanging around the streets of La Granja or to find women occupying the sport field in the center of the village to play volleyball.

Furthermore, among some of the interviewed women, it has been found that higher spatial mobility is associated with growing economic independence. Comparing the data of the national household surveys carried out by the National Institute of Statistics and Informatics in 1999 and 2009, Fuertes and Velazco (2013) analyze the changes in the households' composition in the context of a decade of economic expansion in the country. Among other findings, the authors observe that the increase in households with a female head has been notable, moving from 23.3% in 1993 to 28.5% in 2007. The authors associate this feature of the households' transformation in the country with high female autonomy and empowerment, as a result of the effects of the internal war and migration processes that pulled men out of the homes. But they also relate the changes to greater access for women to education and labor markets. Certainly, more educated women and with higher employment opportunities are more economically capable of coping with separation or to decide to stay single.

Although it would have future consequences for them, the afore-mentioned increase in physical and economic autonomy seems more a result of enhanced economic opportunities than changes in the social and cultural configuration of gender relations and male hegemony in the country. A comparative analysis conducted by the Pan American Health organization among 12 countries—ten of Latin America and two of the Caribbean—revealed that 40% of women in Peru declared that, at some point in their lives, they had suffered physical or sexual violence from their partners or ex-partners. Tied with Colombia, Peru had the second highest percentage in the list (CEPAL 2014, p. 40). In addition, the country exhibits some of the highest numbers of female homicides in the region; of the seven countries in Latin America with available data, Peru was first, with 83 cases of women murdered by their partners or ex-partners (CEPAL 2014, p. 40). Among the studied families, it is important to note that two women who exhibit relatively high levels of physical and economic autonomy experience a tense re-lationship with their partners. During the fieldwork, a young woman who was born in the village but grew up on the coast (and has later returned to La Granja to work in the restaurant of a relative), stated that: "I do not like to hang around with boys from La Granja, they are very male chauvinist and they like to hit women". Mining develop-ment in La Granja, with the opening up of employment and economic channels, has considerably increased social and spatial mobility for many women. This higher mobility and freedom for women—which is associated with stronger city connections and urban life—challenges existing arrangements of relations between genders and hegemonic practices and discourses of male control over women's lives and au-tonomy, a challenge that on many occasions is met with resistance and open gender violence.

Note

1 While the *ronda campesina* of La Granja comprises around 230 men, the *ronda de mujeres* constitutes no more than a third.

7 Representations

As far as space is a social product, cultural geographers have noted that dominant groups look to impose their particular representations over space—including their aesthetic taste—in alignment with their particular class, ethnic, religious, or gender interests (Cosgrove 1985). However, it would be misleading to assume that the elites are the only groups seeking to consecrate their spatial reading, that their representations are met without resistance or that they always succeed. Indeed, various classes and groups compete to impose their views. Richard Peet (1996), for instance, has analyzed how diverse allies in contemporary Massachusetts dispute representations of the local past in relation to their current political, class, and racial struggles. In addition, although dominant classes construct hegemonic representations, subaltern representations are always present and the latter resist as well as accept, adapt, and modify the dominant ones (Lefebvre 1991). Moreover, as Offen (2003) has argued in the case of the Miskitu of Nicaragua, indigenous groups have turned representations of space and landscape into ideational resources in their fight against transnational firms over the control of territory and the exploitation of natural resources. In the context of a proposed gold mining project in the Peruvian coastal valley of Tambogrande, farmers contrasted large-scale images of homogenous and normalized environments without people with local representations that dramatize daily life (Castillo 2006). Indeed, representations of daily life have become at the center of many of the currents struggles against the colonization of space in late capitalism (Lefebvre 2002).

People's representations of space include the productive landscape but also their appropriation through daily life routines. In this regard, gender and age are central components in the construction of La Granja's spatial representations. In addition, people's representations of the past are elaborated vis-à-vis current tensions and negotiations with the mining project. In parallel, their representations of the future express their desires as well as much of their current anxieties over La Granja as a social space.

7.1 The past

For most of the villagers, the beginning of mining development in the region is an essential milestone that marks their reading of La Granja's past. Certainly, past representations of La Granja's social space are arranged between the time before the arrival of Cambior in the area and after it. The establishment of this watershed in the locality's history is perfectly understandable due to the great and traumatic effects that Cambior's displacement caused for most of the families. However, there must be an appreciation of how current processes in La Granja are experienced and narrated, with Rio Tinto's mining development project as background. In this regard, it is interesting to note that, in a general sense, there are two major narratives depicting La Granja in the years prior to the arrival of Cambior.

One of the narratives represents La Granja as a backward space, a sort of sad and isolated place with few opportunities for individual development. For instance, a middle-aged local villager describes his life in the locality before the arrival of Cambior:

> At the time I was a student, the houses were scattered and untidy. The only street was unpaved, there was only one small health center and there was no sports field. The children walked barefoot while the adults just wanted to have animals to sell and to buy more. They did not invest in their children; they just wanted to be millionaires. Many times, I left the school to graze the cattle; my parents did not really push me to study. My brothers, sisters and I were single, worked for our father. Everybody worked in the farm.

The villager represents La Granja as an agrarian space, relatively isolated with minor urban and economic development. In addition, he highlights the family-oriented and patriarchal traits of La Granja's society as well as the secondary importance that education played among family's priorities. Another middle-aged villager elaborates on a similar representation of the locality before mining development in the area:

> The families worked in agriculture and cattle raising, and production was mainly for their own consumption. The houses were built of adobe and education was quite limited; the young people had

to go Chiclayo to continue their studies. My family was the only one that traded cattle but there very few businesses in the town. Women had no opportunities for wage jobs. The selling of granadilla and beans was the main business and the people sold them in Querocoto. There was not transportation, so the people had to go by horse to Querocoto. It was very difficult to travel; one spent the whole day travelling.

Once more, La Granja is represented as an agrarian space lacking communications. Limited employment and educational opportunities forced the young people to migrate to coastal cities. It is an image of hard work and little future expectation of change, especially for young women. It is worth noting that both narratives come from relatively young adults, who are currently engaged in local businesses providing services to the mining project. For them, amidst rumors that the company will suspend the project because of financial considerations, La Granja's past does not represent a desirable future.

In addition, it is no surprise that those people who left the area years before mining operations offer a somber representation of their locality. This dull picture of scarcity works as the background against which their tough decisions become meaningful. A middle-age farmer born in El Verde who migrated with his family to the coast explains that:

Ten years ago, El Verde was very poor and there was only maize for personal consumption. There was not enough land to sell in the markets. The sow was between October and November and the harvest between June and July. However, the production was not enough for the whole year, it finished in December. After December, we suffered from hunger; there was much poverty.

Many of the residents in Querocoto share a similar vision. After all, they are more likely to reap the benefits of mining development without suffering its negative effects. A merchant, born in a rural village who then later moved to the locality, explains that:

Querocoto, despite being the district capital, was a small place; the houses were not built of brick and cement. The transport was difficult and slow, there was only one bus company and there were many risks on the highway because there was no road maintenance. To bring basic goods and to sell farming products from the area, the merchants had to go to Yanacuna, a stop in the highway between Chota and Cutervo, and then take a bus or a truck to

Chiclayo. The people here had no vision for advancement in education and very few people came out from Querocoto to study in Chiclayo or Trujillo. [...] Between 1988 and 1992, there was an increase in cattle thefts and robberies to homes. The cattle rustlers went to the farmers' house and asked them for their plow. Crime was unstoppable; people were afraid.

However, many other residents create different images and contest representations of La Granja as a backward space. Among the interviewed people, it is the seniors and members of the resistant families who create a more positive representation of La Granja's past. For instance, one of the oldest interviewed villagers states that:

> In the time of the *hacendados*, they organized celebrations where there were bullfights. They were good people. The *hacendado* "Uvita" Arrascue treated us well [...] Before, there were trees full of fruits at the border of the road and you could go with a basket to pick them up. I had granadilla trees, the harvests were abundant and I sold them in La Granja. People from different localities, from Pariamarca and Quipayuc, arrived to La Granja's Sunday market. There people wholesaled hens, oranges, limes, coffee to be transported to Chiclayo. People also sold cows and pigs for the slaughterhouse in the city. In those years, people trusted each other, there was more peace in the village, and the cattle were never stolen. I liked to drink, talk and smoke with my friends.

This narrative idealizes the *hacienda* times, through images of a bucolic farming life and the portrayal of a particular sense of social peace and male socialization. This representation, however, conceals certain forms of violence.

The interviewed members of the resistant families share similar representations of La Granja's past, which emphasize features of good weather, familiarity and comradeship, shared culture, collective work, and solidarity. One local resident expresses that:

> [...] the soil was very productive, fertilizers were not necessary. In addition, there were bigger plots of land where the cattle grazed in an open way. The road ended in Huambos, where the people had to go to get basic goods from the coast; nevertheless not much was needed because people did not consume rice but manioc, nor refined sugar, people had their own sugar mills, prepared sugar-cane liquor and dark brown sugar. [...] In the village, all of

us were family related and thus we knew our traditions. I liked
that time because a cooperative system existed, and we worked in
a communal way. When we had to build something, we immedi-
ately started to work without waiting for a salary. There was more
peace in those years; the people left the working tools in the fields
and nobody took them.

A woman from one of the resistant families describes that long ago:
"...there were justice and union in La Granja; the people worked together as volunteers; men and women worked in the road maintenance.
Everything changed with Cambior".

Different from the elders' depictions, for the resistant villagers the
construction of idealized representations of La Granja's past is central to their identity. Their actions during Cambior's land acquisition
processes take on a meaning when contrasted with a peaceful and co-
pious previous agrarian life. Their decision not to sell their land to
Cambior—which in some cases was the result of a delaying strategy
for seeking a higher price that was never ultimately offered—could
thus be reinterpreted as a heroic resistance to a mining company, a
resistance that enabled the survival of the village. In this narrative,
the resistant farmers regard those who sold their land as traitors. As a
villager that moved away from La Granja recalls: "...the people who
stayed in the locality ignored the ones that moved, they thought that
we had betrayed them".

Of course, these opposing representations are a generalization I
have created to offer sharper contrasts. The reality is more complex
and nuanced; and somehow, contradictory views tint individual and
collective representations of La Granja. Therefore, on the one hand,
members of the resistant families also mention some negative features
of the village before the beginning of the mining project development.
For instance, one of the interviewed resistant villagers declares that:

> We lived a bit away from the culture, without roads and without
> fast communication. People easily died because there was not a
> health center in the village, and it took too long to get to the city.

On the other hand, even though a middle-aged resident describes La
Granja's past as a place of lacking opportunities and says that he does
not want to return to farming activities, he concedes that "...the people were more united, everybody fought the *hacendado*". It is interesting to note that this type of narrative depicts a social landscape with
clear positions against which to take a stand and, therefore, it allows

the emergence of a strong sense of solidarity and collective action. Nowadays, the situation is more complex and the frontier between "enemies" and "allies" are blurred. Certainly, mining development has created opportunities for some families; nevertheless, others have been excluded from those benefits. To some extent, this situation has fragmented the collectivity. Some researchers (Damonte 2012) have pointed out the political effects of large-scale mining projects on local institutions. Because local actors will tend to fight over the resources offered by extractive projects and because the latter will tend to privilege and engage with some groups at the expense of other ones, mining development will exacerbate fragmentation and social conflict—among the localities and within them. Reading the data from La Granja's case, however, this statement needs to be nuanced. First, because current mining development has strengthened the *ronda campesina*—the most important institution for local governance—instead of weakening it. Second, the mechanisms for accessing economic benefits do not necessarily imply an individual atomization. The access to the temporary employment program is mediated through the collective action of the *ronda campesina* and, maybe more significantly, the creation and implementation of local businesses implies the bolstering of extended family networks.

Moreover, representations of the past as a time of solidarity and abundant farming production should be understood in the context of present-day tensions over mining development. In this way, the same villagers that draw a picture of La Granja's past as being a backward place, when confronted with the changes prompted by the mining project, point out two issues: the loss of cultural identity and social bonds due to the arrival of outsiders, and land and water contamination. A local resident states that because of the arrival of new people in the locality, "...customs, saint festivals, religious celebrations and baptism ceremonies are being lost". In addition, an immigrant residing many years in La Granja indicates:

> The land was very productive, you sow something and everything was harvested, there were large variety of products. Likewise, people were nice, they invite you to have some meals anytime, they never sold you food, and it was given for free. The mines have spoiled everything and the land has lost productivity because of the pollution.

Many leftist and environmentalist groups opposed to mining activities use similar discourses that question experiences of social change.

Against a triumphal and hegemonic vision of modernity developed by representatives of the industry and central government bureaucrats, these groups support a conservative recreation of an agrarian arcadia, a lost rural Eden. The discourse developed by Marco Arana—a well-known Congressman and regional leader opposed to mining activities in Cajamarca, especially against the presence of Newmont's Yanacocha gold project—is an example of this nostalgic recreation of a past social landscape:

> Our city began to grow exponentially and was packed with foreign people, the doors of our houses would not stay open anymore and the bikes of the kids would never be safe on the sidewalks again. The beautiful valley began to disappear to give way to mega warehouses, business that range from the sale of heavy machinery to the illegal brothel; the rivers where we use to splash when we were kids and where we ended the day catching *charcoquitas* [small river fish] and catfishes no longer exist. This is not only because the mine has run out the natural sources and left the Mashcón and Chonta rivers dying, but also because the gas stations, the *mototaxis* [tuk tuks or cabs propelled by a motorbike], and the local government, which do not adequately manage the sewage. All of them have contributed to provide everyday a new *coup de grâce* to the little water left in those rivers. And if you are in a hurry, don't even think about going into the downtown of the still small city during the rush hour or rather go at that time [if you want] to breathe your dose of modernity, of polluted air in the style of Lima's Parque Universitario [a popular and crowded park in Lima's downtown].
>
> (Translated from Arana 2011)

This discourse emphasizes and idealizes the dimensions of social cohesion, familiarity, solidarity, closeness to nature, and abundance. However, it hides other aspects, such as poverty, gender violence, patriarchy, low social mobility, and lack of job opportunities, among others. I am not suggesting that transformations brought about through mining development have solved these problems, but they have altered established ways of representing society. The insight I would like to emphasize is that representations of past social landscapes cannot be taken for granted but constitute symbolic resources of present struggles for the access and control of natural resources as well as the definition of mining development and appropriation of its benefits.

7.2 The present

In a double-movement relation, present-day interests shape the construction of past narratives as well and later influence current representations. In this sense, people directly engaged with mining-related activities and benefiting from the project, those who migrated out of the region fleeing poverty conditions, recent immigrants that have arrived seeking job opportunities, and young people will tend to develop more optimistic visions of the present and will contrast them with relatively harsh representations of the past.

For instance, the male informant of the Resistant family 1 had temporarily worked at the mining project with Cambior, BHP Billiton and, at different times, with Rio Tinto. In addition, as a builder, he had benefited from recent housing demand in the village. He expresses an optimistic view of present-day life in the locality. As a result, he highlights positive performance and effects of current mining development around environmental aspects:

> Rio Tinto demands many local workers and provides high safety conditions. Any accident would generate discontent within the village, for that reason the company train local people; the company seeks to avoid accidents. In addition, the company seeks environmental safety and, thus, it builds pits to deposit polluted water.

And especially on economic and social issues:

> Nowadays the houses are increasingly tidy. There is higher economic activity in the village because the demand has increased as people have a fixed income, subsequently; there are more local businesses and stores. The surrounding localities bring more farming products to the Sunday market. [...] These days, parents think in investing in their children's education, they have realized that when the company arrives they do not have enough training to get a better salary.

To the new economic and employment opportunities, a member of one of the returnee families adds positive improvements in housing, public infrastructure, and public services as well as general safety:

> Nowadays the houses are more modern; they are made of brick and cement and are more in number. We have water service at our homes over the past ten years and sewage system for four years.

The roads are in good condition. There are schoolteachers for the three levels [initial, primary and secondary]; before there wasn't, the students entered older at the school because there was neither kindergarten nor an initial level. There is a health center now, before one had to travel to Querocoto to get attention.

It is no surprise that members of opportunistic families, those who arrived in the locality in search of employment and economic opportunities, narrate visions of confidence of present La Granja. One of these residents considers that the village is now a better place to live than it was before because:

> Due to the mine, everybody has a job and you can sell what you have, that is the reason of the street market. There is the Sunday market, when people from Vista Alegre, Querocoto, Mitobamba come to buy and sell. There are restaurants, convenient stores, billiard rooms and cockfights on Sundays. Moreover, with the night patrolling, there are no robberies and one can walk even late at night.

Those who have moved out of the village also create positive representations of La Granja. In general, they construct images of progress linked to economic growth and material improvement, which lead to better living: "La Granja is a better place because everybody has good houses, not like before when the houses were built of mud and straw; people are wise and build pretty houses to get a better price from the mining company". Or, as another emigrant states:

> Everything has changed, young people are calm, happy because there are jobs, and the mood has changed. There are more houses in the town, more people, more happiness and the youth practice sports. There are restaurants, more cars, there is electricity and television. There are more businesses and if there are more people, everything is better.

Some immigrants even consider the physical environment to be improving. The male informant of the Migrant family 1 expresses: "there are many trees now because people are reforesting". Indeed, there is currently less of a threat to the forests from farming activities because of the availability of nonfarming paid options.

Nevertheless, many of these immigrants have moved to coastal regions and, although they maintain strong local networks, they prefer to continue their lives there. This may be because coastal cities represent a modern lifestyle. The male informant of the Migrant family 2 explains: "I have furniture and electrical appliances for my house in Chiclayo; here [in Chiclayo] you get more variety at a cheaper price" or because the cities are better linked to markets, as the male informant of the Migrant family 4 states:

> On the coast, if you don't have lands you can buy products from other farmers. I usually go to the Mochoqueque market, in Chiclayo, in my own motorbike equipped with a hopper. The trip takes me one day, but when I was [living] in El Verde it took me three days.

As might be expected, members of the regional families—namely, those living in Querocoto—consider the town to be experiencing largely positive changes due to the mining project. There are better roads and a significant improvement in transport facilities, which makes it faster and cheaper to trade products between the village and the coast. The arrival of newcomers and the implementation of project-related investments have fostered a sharp increase in local businesses—mainly in lodging, construction, rental of light trucks, gas stations, and restaurants—and higher competence has produced better services. With more working opportunities, including those for women, and a reduction in unemployment, criminality has decreased. The economic prosperity of the town has translated into more and renewed houses as well as better education, health, and sanitation conditions. In addition, some farmers have introduced technological improvements in their practices, such as the adoption of spray irrigation, crop intensification, and new techniques for growing coffee, and they have been able to break the seasonality of some products in order to avoid overflooding of the market. Activities of the development programs implemented by Rio Tinto's social fund partly explain these farming improvements. However, in contrast to what occurs in the intended area of resettlement, these programs are relatively successful because the people from Querocoto expect to continue living in the locality. In this sense, their investments in productive activities in the area are linked to their present and future representations of the place.

To some extent, as previously stated, these groups of villagers express their perceptions of a frantic and progressive present against a narrative of economic stagnation and physical isolation in the past. As a merchant from Querocoto starkly puts, "in the past, this place was dead". This break with the past would have even produced changes in people's mentality, especially among young people. The male informant of the Resistant family 1 declares: "With the existence of roads, young people become more active, when you have more money your desire to move out of the village increases".

However, the elderly and members of the resistant families tend to construct representations that challenge the previous images of present-day La Granja. Although they acknowledge some positive aspects of the current mining development, such as the opening of local business and employment opportunities, the rise of economic dynamism, and the provision of better public services, they are more likely to highlight the negative factors that the project is imprinting in La Granja's social space. A local resident depicts the changes experienced in the village, explaining that:

> Nowadays, we have the provision of basic services; the village looks more like a formal city where we live in a different way. Job opportunities have increased, especially for men. Nevertheless, these opportunities are only for skilled labor because Rio Tinto makes a selection. Thus, the people that work in the project are mainly from Chiclayo and not from here. With so many outsiders, the trust of old times is lacking, now we have to be careful. In addition, the major part of the people only think about personal benefit, people are used to charity and hope to receive everything from the company. This is because when people ask a request to the local government and the company, the latter answers in a more effective way. The authorities neglect the locality because they consider we have enough support from the mining company.

Another member of a resistant family supports these representations of current La Granja, stating that:

> The population has doubled. [...] The arrival of many outsiders has increased the price of street food; there are two laundry services that employ women and there are two businesses that lease water tanker trucks to the mining project. These businesses provide employment, but mainly to outsiders. In addition, the people do not want to grass their cattle and cultivate the land because they see that business is easier and more profitable.

A relatively old villager complains that nowadays:

> The day's wage for the laborers has increased; they want to earn the same as the mining company's wage. The people that work

for the mining project could go out and travel, they earn enough money while for us, the ones that work in the farm, our land is only to provide us with food. [...]

The above representations of La Granja today raise questions about some of the stated benefits of the project for the local population. These representations indicate major social issues in current debates about the consequences of mining development in rural regions. First, there are the different interests as well the different evaluations by people who have engaged in mining-related activities and those who still mainly depend on agriculture. In a scenario of sharply increasing local prices (Viale & Monge 2012), mining development could seriously affect diverse groups, especially in housing and food provision, and they would raise their voices of discontentment (Vega-Centeno 2011).

A second consideration is the clash between local expectancies of massive and permanent employment and specialized needs of the company and relatively short periods of the labor peaks. The divergence arises from the limited capacities of local economies to provide labor and services to a highly specialized and capital-intensive industry (Mendoza 2011) and the difficulties of building regional productive clusters (Kuramoto 1999). The management of these expectations is central to the adequate development of the project and is often at the core of its social responsibility programs.

A third issue is the marked shift in local migratory and demographic dynamics. In terms of representations, one of the most remarkable consequences is that the localities have moved from places of expulsion to places of reception. Perceptions of mistrust, insecurity, and a dislike of outsiders, and a general sense of a loss of community, are better understood in terms of these changes. There is a sense of fluidity, change, and instability that significantly modifies the quality of the social capital (McLean, Schultz & Steger 2002). This is to say, there has been a move from "bridging" social capital—which links people to social networks and markets—to "bonding" social capital, which leads to more insular behavior as expressed in the hostility of outsiders.[1]

A fourth issue is the political tension between the local governments and the populations within the area of influence of mining projects. One of the points of tension emerges from the provision of public services to the population within the direct influence area of the mining project. People and authorities from surrounding areas consider that the areas adjacent to the project receive large benefits from the company's social responsibility programs, so those populations would not need more attention and the diversion of public funds. Generally, populations within the direct areas of mining project's influence are rural and relatively small; therefore, their political ability to influence the local government's decisions around the allocation of public services and resources is limited (Soria 2014). Abandoned by their local and regional authorities—and often by the central government too—these populations will turn to the mining company for the solution to their demands. The lack of attention of local and regional authorities to the local populations, then, reinforces the formation of the patron–client system between the latter and the mining project, a dependency process that Salas (2010) has acutely analyzed for the case of the San Marcos district, under the area of influence of Antamina's copper project in northern Peruvian Andes. Therefore, a political dependency is created in addition to an economic one.[2] Furthermore, as Soria (2014) has examined for the case of the province of Espinar, Cusco—where a Glencore copper project operates—while transparency efforts from civil society groups have mostly centered their attention upon the company's social responsibility programs, it is the local governments who manage the greater proportion of the financial resources in the territory. The abundance of these resources has fueled significant cases of corruption as well as conflict over the political control of the local governments (Arellano-Yanguas 2011).[3]

Despite these different and sometimes opposing representations of La Granja, the local people are aware that the economic prosperity they are experiencing depends on the mining project. As a local resident wonders, "La Granja now has roads to bring in trade products because of the mining project, but if the company leaves the area, who will keep up maintenance?" Another resident fears the abandonment of the mining project because it "...provides us with employment and we have got used to this kind of work". An emigrant who left the area almost 15 years ago and now lives in a rural area in the coast of Lambayeque expresses an even grimmer vision: "El Verde is a bit better now, but this is just a temporal relief due to the mining-related employment. Agriculture continues in the same bad situation. The highlands have no alternatives; it is a very poor region". Indeed, the whole transformation of the space is linked to mining development; when it stops, it will repeat the cycles of creation and destruction, which are characteristic of capitalist development. Of course, these uncertainties are the main sources of anxiety about the individual and collective future of the *Granjinos*.

7.3 The future

To some extent different from the representations of the past and present, the representations that people present about La Granja's future are less aligned with individual and family interests and positions. In other words, although young people and returnee, opportunistic, and migrant family members emphasize some different aspects from elders and resistant family members, there is a shared sense of uncertainty and anxiety about the future. In addition, also different from the previous representations, people construct less elaborate narratives, maybe precisely as a result of the uncertainty and the difficulty to foresee the future.

Of course, many *Granjinos* imagine a better future life outside the village and on various occasions, their desired future place to live is a relatively calm town in the highlands, in the same region. A local villager declares:

> I imagine myself living in Chota; it is a strategic place, quiet, with good security and good relation among the neighbors. People are progressive, with future alternatives to start a business. In addition, I am from Chota, I have friends and relatives there, I know the situation of my town, and there is not so much selfishness. I am willing to go because of my kids. I want to buy a computer for them and a car to move my brother [who is physically handicapped]. In the future, I would like to have an excursion house in the country, not built of brick and cement, a house where I could have my own world, with a children's playground, a little lake where I can raise ducks, with trees, cattle and big enough to bring the family together. My dad had a project like that. I also want to improve pastures and crops and to reactivate beekeeping and the breeding of small animals; it is a profitable business. I would die on the farm.

Nevertheless, not everybody agrees with these images of a romanticized rural landscape untouched by modernity. As a male migrant informant from La Granja who currently lives in Querocoto explains, "Sometimes I think to sell everything and move back to Chota, my hometown. However, everything has also changed there; the town has grown".

Indeed, for a good number of the *Granjinos*, migration to a coastal city—mainly Chiclayo—is the preferred and envisioned future, even if the mining company does not go ahead with the project:

> In the future, I imagine that we will be living in Chiclayo because of the education of our children. A good zone would be

Pimentel or suburbs like La Victoria or Primavera; they have paved roads and walkways, they are not slums and we can save on transportation costs because they are close to the universities. I would like a two-story house, with patio, carpark and all the services. I would like to have a refrigerator and a car. We would like to buy farming land to cultivate rice or maize. In addition, if the project continues, I would lease heavy machinery to the company.

Somewhat different from the rural ideals of the previous informants, the latter narratives emphasize the perceived benefits of urban life: higher education opportunities, better public services, and expectations of urban and modern comforts. Nonetheless, these residents also wish to keep ties with farming activities. These representations do not imply a sharp break between country and the city, but a continuation of their previous lives and in complement to them. Read in the context of the large urban-country inequalities existing in the country, the statements are an argument for continuing with a farming lifestyle without suffering the lack of opportunities. In addition, there is also no break between the present and the future in many of the local narratives. People not only want to continue with some forms of improved agriculture, but also want to continue the link with the mining project for the rest of their lives. A male informant for the Resistant family 1 argues: "The solution is that Rio Tinto supports us to build local businesses to continue servicing the project. I want the project to give me work until I die or until I can no longer work". In this sense, the expectation is that in exchange for a permanent asset—land—the mining project should provide permanent benefits to the local population, not just discrete compensation packages. They expect, in other words, one livelihood in exchange for another.

Once again, the residents of Querocoto construct the most optimistic image:

There will be large benefits if the mine is constructed. Just the work of the Social Fund will last two years; it is not good to oppose the mine. Querocoto will be a strategic economic center and we need to be prepared. Investors will come to open businesses and hostels. For that reason, I have planned to buy a terrain on the outskirts to build a hostel. Previously there was no economic movement. [...] Construction will increase in Querocoto because people from far away will arrive to buy and sell products; there will be more houses; a peripheral road is already under construction in an area without houses and the town will grow. I would like that those who have not had opportunities now have a chance to get a house and educate their children, so the children could have different jobs from their parents. The company could stop operations but if the children are educated, they can work anywhere.

A local woman, who owns a small business in Querocoto, declares that if the project continues: "the work for women will increase because they are more skillful than men. Furthermore, there would be less crime because people are making a great effort in training themselves in order to access the employment opportunities".

Along with these promising views of their own future, people imagine somber representations of La Granja as space:

If the mining project is developed, La Granja will be a desert because of the construction of an open pit and the removal of land. This land will belong to nobody in the next forty years of exploitation. La Granja will disappear.

It is interesting to note that many interviewed people imagine La Granja's future as a desert, an empty place without life, or, even worse, a polluted place of desolation:

I don't know what a project looks like; I have never been in a mine, but certainly, there will be no plants, the mine destroys everything; copper damages the land. The wind will be more polluted. People will move out and will have to eat another kind of food. The houses will disappear.

Moreover, social relations will be disrupted; foreign people will arrive, the local ones will be displaced, and families will be split and moved to different places. The uncertainty about the future of the project and people's own lives, which is source of sadness and nostalgia, produces a sense of loss of place, a lifestyle, and social bonds, as a local villager dramatically expresses:

When the machinery enters and the land movement starts, the place will be devastated. A large camp full of foreign people will be installed. There will be environmental contamination, with noise and smoke, there will no longer be water and vegetation. It will be sad to see the place where my son was born. The school will be buried. I will see La Granja as the end of what it was.

An immigrant who has established himself in La Granja more than a decade ago mentions:

> If the project goes ahead, the company will expulse the people, only the company's people will remain to do whatever they want. The village will disappear and foreign people will arrive, we will never come back. There will be nothing here; it will be just a desert. I would not like to move because I am fond of this place. I will not find in other places the same that exists here in La Granja.

One local merchant from Querocoto mentioned that with the project operating: "Many investors will arrive; there will be more competence and they could take our clients away. This is a fear we have. There will also be an increase in the population and sexual freedom; that is scarier".

A local woman from a resistant family depicts one of the darkest scenes:

> It will be a disaster if the company develops its project, the lands will be spoiled, and the water will be polluted. We see how the mining companies do that. Nevertheless, nobody wants to sell their land; for that reason, the company has called the police; perhaps they will kill us. [...] There will not be working opportunities, because they want professionals. What they promise is a trick until they expulse us from here. They will not consider the local suppliers. If they are doing awful things now, could you imagine what they will do later when they become the owners?

Faced with a dull future—represented as a physical, social, and moral disorder caused by mineral extraction—many people consider the possibility of moving especially to large cities on the coast, as an opportunity, but also a risk:

> We are afraid to be robbed when moving to the coast. Our fellows know us, and they can spread the news that we are arriving to the city with money from the sale of our properties to the mine. In addition, people in Chiclayo are not welcoming.

Furthermore, some people are wary of the mining company's promises of future support and fairness. A local villager is worried about being abandoned by the company and declares: "The company mentions that we will be supported for one and half years; but that time is short, considering there are elderly people. The mining company will expulse us in different places". Another villager reinforces this distrust: "I don't think Rio Tinto will fulfill its promises to support the resettled families. The company will not be interested in them when they have already moved".

The combination of these factors—reluctance, uncertainty, and mistrust—causes some people to doubt the feasibility of the project itself. For example, a local resident believes that the mining project will not continue because "...according to the company's policy it is enough that someone is not in agreement with the project [and therefore the company must] abandon it. This seems unbelievable but they [Rio Tinto's employees] say that in their meetings". Other villagers consider that they will not move, and they will resist the implementation of the project; they want to continue living in La Granja:

> I would like to live in peace. In the coast, life is too busy and people age fast. I will continue with my same business [lodging and catering] and foresting. I also would like to implement a coffee and granadilla project. I do not have an ideal place where to live; I want to continue living here in La Granja and I will not sell my properties to the mine.

In this context, some families have developed a strategy for benefiting from the mining project without selling the land and thereby allowing the project to be developed. As a local resident explains, "let them [the employees from the mining project] come here to provide jobs until they get tired".

7.4 Representations of space and construction of place

Through their representations and memories, people appropriate otherwise anonymous space and convert it into their own vivid and unique place (Cresswell 2004). However, the appropriation of space and construction of place is not a mere act of individual memory. This process of remembrance is anchored in the collective struggles over place and territory, which, in La Granja's case, have moved from being a struggle against the *hacendado* to a conflict relationship with the mining project and its different operators. Certainly, the mining project works as both a temporal and geographical landmark from which local people produce their representations of the past, present, and future of the village. In La Granja, mining is a central element of the "sedimentary landscape" (Moore 1998) and the recreation of its historical legacy is central to current local micropolitics.

For many local people, the changes brought about by the mining project are represented and experienced as a tension between, on the one hand, future economic and education opportunities and social and spatial mobility and, on the other hand, the destruction of place. The destruction of La Granja as a place is conceptualized in three dimensions. The first one highlights environmental aspects related to perceptions of contamination of the soil, water, and air. The second is a social dimension that points out the breaking of social bonds, the separation of families, the loss of solidarity and cultural traditions, and the relaxation of morals (especially in terms of the women's sexual behavior). This is a dimension that stresses social anomia, which is produced by the changes in lifestyles from the country to the city (Harvey 1985). The third feature comprises symbolic elements as far it refers to the vanishing images of homeland.

Along the two axes—economic benefits and social and physical mobility, and loss of place—people position themselves in complex, fluid, and sometimes contradictory ways according to factors of family and personal history of engagement with the mining project, age, and sex. Therefore, older people, men, and those positioning themselves as resistant to the mining project will be likely to produce a past rural arcadia of peace and harmony, bucolic agrarian life, abundant production, solid social bonds, solidarity, shared values, and closeness to a pristine nature. They will use these representations as a background against which to contrast what they regard as an unruly present. At the same time, when imagining places for future living, they will create spatial utopias, a recreation of the desired La Granja in other locations. Places such as Chota emerge as idealized rural landscapes of bucolic relations of communal work and solidarity. It is worth noting that, as in the case of the male informant's narrative of the Opportunistic family 2, the landscape is linked to the figure and memories of his father. As it has been noted in the chapter on La Granja's history, Chota is the place of origin of many *Granjinos*, and moving back to that warm and fertile Andean province is not merely an individual return to the homeland. It is the symbolic return to a rural place where there is no mining development, a return to the lost arcadia of solidarity, selflessness, and contact with nature. Certainly, it is a sort of recreation of "the land without evil" that many indigenous populations sought in their desperate escape from the ethnocidal effects of the colonial orders in South America (Clastres 1989). In any case, the constant moving back and forward of the mining project has also produced uncertainty and anxiety about their future lives for many *Granjinos*. Albrecht (2005) has coined the neologism *solastalgia* to refer to the stress that environmental change, such as the one produced by large-scale mining operations, causes people, and the sense of a lack of control over global and external forces. In the examined case, despite there being no significant environmental changes, given that the project has not been developed yet, people are likely to manifest their social distress using an environmental reading. The environment becomes the arena for social struggles over space and economic benefits.

Other members of La Granja's society, however, contest the previous representations. Women especially refuse to accept the narrative of La Granja's past as a heroic fight, first against the *hacendado*, and then against Cambior. They reject a vision of a peaceful patriarchal society, which hides social conflicts, hustling, heavy drinking, and, more significantly, domestic violence. Therefore, for some elderly female interviewees, La Granja's past represents a closed space where their movements were strictly limited to their homes. For these women, La Granja's old days depict a space of male fights, heavy alcohol consumption, jealousy, and insecurity:

> In the parties, men got drunk; they fought and killed each other. I did not like to go out of my home [...]; I saw the fights from far away and I was scared of the killing. My husband was always watching over me when he was drunk.

Furthermore, many women have struggled for the breaking of social barriers and salaried work. In this sense, the mining project has meant an opportunity for change from which many women have directly benefited. Therefore, it is little wonder that women construct less idealized images of the past and view current conditions in the village in a more positive light. In addition, significantly more often than men, women tend to link their demands to the benefit of their children and family—for instance, employment opportunities or better health and education services instead of cash compensation or productive projects—rather than to themselves. Similarly, a local woman indicates that she likes more present-day La Granja more than before because "there are jobs for our children, otherwise they would have to move out to other places and the family would break". This situation is consistent with findings in other mining regions (Soria 2012). Maybe because women's political participation in the area is notably limited, they tend to support or oppose the development of the mining project due to interests directly related to their livelihoods and less to ideological positions.

Along with sex, age also shapes representations of social space. Thus, for instance, adult women tend to portray the city—namely

Chiclayo—as a space of danger and insecurity that restrains their physical mobility. These women perceive the city as being a sort of maze, a labyrinth where it is difficult to read its signs. On the contrary, although young women recognize the city's lack of security, they emphasize the different choices it offers, the possibility of finding paid work, making friends, establishing relationships among peers and strolling around the streets and shopping malls without the family constraints they experience in La Granja. Younger generations mostly perceive rural localities as being backward places, where there is nothing to do except work around the mining project; as a woman in her early thirties points out:

> I did not like Chiclayo at the beginning; the streets seemed all the same to me and I would get lost. Then I made friends at work and we strolled along the streets. I do not miss La Granja; every time I go, I get bored. I like how it is now more though; before there was nothing, not even jobs. Only old people miss farming, only old people like how La Granja was in the past.

Nevertheless, it would be misleading to think there is a sharp and clear distinction between men and women's representations of La Granja. Not all men idealize the past and not all women contest the hegemonic vision of La Granja's past. Certainly, some women do not even question the old patriarchal system but miss the stability it provided. For them, the changes that mining development has brought into the area have created social disorder, which is verbalized in terms of fears arising from untamed sexual behavior: prostitution, licentious lifestyles, marital infidelity, or separation and divorce. A local woman describes a direct link between the mining project, access to money, and marital disruption: "Women leave their husbands, they are dirty women; I do not like that. With the arrival of the company, men from other places come; women then establish relationships with married men; they are lovers for money".

Indeed, this is not necessarily an opposition between men and women; it is about how gender relations—which include practices, roles, and discourses—are conceived and about the moral statements that local people express regarding how these relations should work. Therefore, in some cases, both men and women assume strong positions against what they perceive to be an attack on the established social and moral order. In this sense, representing the spearhead of modernizing processes in the local area, some people consider that mining development does not only destroy the material landscape, but that it also erases the prevailing social relations among villagers, and among men and women; it is "…the end of what it was", as expressed by a male resident. For this reason, the conservative agrarian utopias that some people create reflect a desire to return to an idealized place prior to mining development, not merely in the sense of contact with nature but also in the sense of a patriarchal society of collective work, solidarity, benevolent authority, and family-oriented values.

In a provocative and incisive essay, Guillermo Nugent (2012) analyses the distinction between "destiny" and "project" in the context of the exclusive and hierarchical class and ethnic order of Lima's society in the 1960s. While, from the perspective of the privileged white minority, subaltern groups (Indigenous, African descendants, and *mestizos*) should simply follow their "destiny" and become farmers, bus drivers, or minor public servants, for instance, without questioning the status quo, the members of the fortunate class should follow their own personal "project" and, for instance, become writers, artists, or whatever they choose. Through their future dreams—whether to become professionals or merchants and live in modern houses in the city, or practice technical farming in pleasant rural communities—and their everyday actions, the men and women from La Granja build projects and refuse to be categorized in terms of an expected shared destiny.

Notes

1 David Brereton, personal communication.
2 As a local informant from an opportunistic family stated: "People have been merely satisfied working in the mine and have stopped farming. Everybody has wasted their income consuming unimportant things, there are not savings".
3 The majority of these resources comes from the distribution at regional and local level of 50% of the income taxes paid by mining companies to the central government, a system designed under the New Extractive Industry Strategy (Arellano-Yanguas 2011).

8 By way of conclusion
What is new under the sun?

Although the material analyzed in this study is taken from a specific case, as befits the ethnographic project approach, the aim has not only been to provide a detailed understanding of the particular dynamics that produce La Granja's social space, but also aspire to shed new light on how current rural societies around the world are experiencing the "friction" (Tsing 2005) that global capitalism unleashes—through the medium of large-scale mining development.

La Granja is a valuable case study from several perspectives. First, there has been more than four decades of exploration and development activities in the area, involving diverse companies operating under diverse international, national, and corporate governance regimes. Most other mining-related case studies focus on the changes a single mining company has brought to local families in a relatively short period (Bury 2004; Salas 2008; Gil 2009; Himley 2010; Li 2015). The La Granja case allows the analysis of cyclic dynamics of household strategies in the long term and also enables the approaches of different companies to be compared. The fact that a mine has not been yet constructed is a second distinctive feature. While other studies stress environmental drivers (Bury 2004; Bebbington & Williams 2008; Bury 2011; Himley 2014; Li 2015), La Granja shows that significant local social and economic change can occur even in the absence of major environmental impacts. There is extensive literature on mining-related conflicts in which environmental concerns typically play a key role (Bury & Kolff 2002; Bebbington 2007; De Echave et al. 2009; Arellano-Yanguas 2011; Li 2015; Schorr & Dietz 2018). La Granja, by contrast, stands out for the absence of considerable and open conflict between local populations and the mining company. This facilitates the examination of local strategies other than resistance and violence. We may recognize, though, the protracted uncertainty around the project. This is a challenge that *Granjinos*—as many communities in areas that have "prospective" resources—will have to contend for many years. Finally, La Granja differs from many other Andean communities that have been studied in Peru (Salas 2008; Gil 2009; Himley 2010; Burneo & Chaparro 2011), in that it does not involve a *campesino* community, with the collective forms of land ownership and land governance that characterize much of Peru. Instead, it provides an example of a land ownership regime individually managed by each family and highlights the speed and ease of land transactions in the context of mining development. These distinctive characteristics challenge some of the current accounts of the social effects of mining development and highlight the importance of historical depth, stages in the mine cycle, and land ownership regimes for the understanding of complex and contradictory processes.

8.1 Changes over land access and land value

Issues over land access have been a central theme in La Granja's contemporary history. This is not an example of a homogenous community enjoying stable livelihoods being suddenly interrupted by the arrival of a mining project. Instead, La Granja's history is marked by a cycle of struggles concerning land that have involved multiple actors: landlords, tenants, the state, mining companies, local families with diverse interests and socioeconomic conditions, and outsiders. Mineral development is only the most recent of these cycles of struggle over the appropriation for accessing space in the village and will probably not be the last. On the one hand, mining has led to dispossession (during Cambior's land acquisition plan) and generated high levels of uncertainty (with the ups and downs of the project during Rio Tinto's management). On the other hand, it has also created some new forms of security by pressuring state agencies to accelerate land titling within the local families; albeit this was done mainly to make it easier for global forces to access and exploit natural resources.

Once a large mineral deposit is identified, the value of land changes from a production factor (to farm or raise livestock, for instance) to a "deposit of value" (Glave 2008). In this way, mining development promotes the formation of a local land market, which is not tied to increases in farming production and productivity. This is reflected in La Granja, where families have seen a rise in their incomes, not because of farming activities but due to nonfarming activities. These increased values have not only been reflected in direct transactions between mining companies and local families, but also in transactions between families, in the context of high expectations of future economic benefits to be derived from potential negotiations with mining companies. This is a common, but underresearched, phenomenon in mining areas in Peru.

The classic literature on rural modernization in Western contexts has emphasized a homegrown market integration, as agrarian production rises, and the surplus is sent to satisfy a growing demand in industrialized cities. As Jacobsen (1993) has noted for the case of Puno, the modernization process that current mining development is promoting among rural societies in the country is not based on the increase of farming production, quite the contrary. The greater economic interaction between rural and urban spaces is due to the increase of paid labor in nonfarming activities and the consumption of external goods. Because the latter is a short-term and exogenous factor, the resulting market integration is weak.

Local families have not moved to cities in a permanent form to be incorporated in the industrial sector, although as a result of the strategy of double residence, some members of extended families now live in the city seeking education and temporal employment in the service and informal sectors. In the context of mining development and economic growth, local families extract resources from rural areas to the cities. The direction of capital flows is thus quite different from that stated in the standard literature on remittances (Altamirano 2010). The capital follows the desires, expectations, and plans of the local families, and representations play a major role in shaping their desires and actions. When mining activities and national growth decline, many members of the families return to farming activities, which provide a refuge of sorts. Using these social networks and productive strategies across economic cycles, local families have established fluid relations between urban and rural spaces.

The rural world is now less dependent on agriculture and is becoming more diversified; in many areas of Peru, mining development is a very important factor explaining this shift. This is not necessarily because of environmental impacts that would negatively affect rural livelihoods. For the local families of La Granja, socioeconomic factors have been the major drivers for economic change and diversification. Wage employment, local business opportunities, and the construction of a road connecting La Granja with Querocoto and the city of Chiclayo on the coast—a road that altered much of the previous trading system—have been instrumental in the transformation of the local economy and the redefinition of the village from a place of expulsion to a place of reception. Mining is shaping a reconfiguration of regional axes with the displacement of previous farming and trading centers to new nodes of capital accumulation.

In brief, the experience of La Granja provides an example of how transnational companies and national institutions prompt transformations within regional and local spaces. The case also highlights that the transformations brought about are not an automatic outcome mechanically derived from global forces; on the contrary, the diverse set of sophisticated strategies deployed by the local families reveal a large amount of local agency. Nor is the use of such strategies unique to this one location. Bainton and Banks (2018) have also shown how residents in Melanesian mining regions use land as a social relation that allows or restrains inmigration flows and access to benefits derived from mining activities. As in Melanesia, local organizations and households mediate global processes and significantly influence the speed and direction of the changes.

8.2 Fluidity of rural—urban mobility

Some of the current literature exploring the effects of large-scale mining on rural societies see mining development as causing major emigration flows, because of the displacement of people from their lands and the adverse effect of mining operations on local livelihoods (for instance, Bury 2011). The La Granja case presents a more complex panorama. First, there have been many migratory waves in the region's history which pre-date mining, and which have mainly been a response to poverty. Mining development did contribute to population outflows in the Cambior period when people were displaced from their lands. However, more recently the La Granja project has increased employment and business opportunities, which has led to more people arriving in the area rather than leaving it; albeit this may be a temporary and nonsustainable phenomenon. Second, and perhaps more importantly, the La Granja case demonstrates that migratory flows are neither unidirectional—from the country to the cities—nor permanent. Rather, they are part of elaborate strategies—such as double residence—which allow households to access the housing and education advantages of the cities while retaining rights in the country. The members of the households move back and forth following economic cycles.

In the context of mining development, people employ these strategies more often and use extended kinship to claim local benefits (Castillo & Brereton 2018b). Consequently, local families have developed flexible productive and mobile strategies in a broader area that blurs the distinction between the country and the city. As rural areas become more urbanized (with the spread of television or the consumption of urban styles and goods, for instance), there is also a ruralization of the city. In their occupation of areas outside the region, the *Granjino* immigrants have taken with them their social and cultural capital and adapted it to the new areas (for instance, the creation of urban patrols and associations of fellow villagers in the city). This is a good example of how rural Andean families have reproduced social and cultural strategies in the new urban context, a process that early anthropologists have studied in depth.

In summary, unlike the classic rural–urban migration in the West, the La Granja case demonstrates that urbanization is not a unidirectional transition from rural to urban lives. The emerging pattern is a mixed and fluid one where families use their networks to bridge both spaces. The strategy of double residency has allowed the local families not only to secure their rural properties but also to avoid becoming fully proletarian. In this sense, mining development is contributing to a major trend in the country where people from rural areas are leaving farming activities and farming identity but are not necessarily becoming proletarians (that is, workers in industrial cities who do not own anything but their labor). The practice and expectations of some local villagers—and not only the elderly—of buying farming lands on the coast and continuing to work in agriculture although living in the city must not be regarded simply as the product of nostalgia for the lost rural world. It is also a statement against the separation between working place and living place, which is characteristic of industrial conditions and which people perceive as alienating.

As in many other rural areas in the country, many *Granjinos* have placed their expectations of future in the city. This explains, for instance, the direction of the remittance flows from the country to the cities, where families invest in education and housing. However, the situation of many immigrants in the city is far from ideal. The failed industrialization of Peru has led to the growth of the informal sector. In a context of pervasive and long-standing gaps between rural and urban living standards in the country, it is no surprise that young people seek to migrate to the cities despite the high levels of crime there. Although people see the city as a place of risk as well as opportunity—a place of change and mobility—the lack of security of informal jobs and the poor quality of education demonstrate the strict limits of the urban myth in the country. The *Granjino* families, for whom the labeling of rural or urban makes little sense, use any strategy—through formal or informal means—to build a better life.

8.3 Weak institutions, strong social networks

The historical account of La Granja indicates that a legitimate democratic system with efficient institutions did not replace the collapse of the *hacienda* regime. During the 1980s, with the profound economic crisis and the civil war, the state retreated from vast parts of the hinterland, economy, and society. Informal organizations and activities—such as informal and micro-businesses, self-defense committees, and independent political movements—partly filled the political, economic, and social vacuum. The neoliberal reforms implemented in the early 1990s, with the deregulation of the economy and further reduction of the state apparatus, accentuated the informality of the Peruvian society. Moreover, after Fujimori's coup d'état and the dissolution of parliament, the government coopted the judicial system and utilized extended mechanisms of corruption and a patron–client system to obtain political control. In La Granja, local families experienced how the central government backed the mining project and put pressure on them to sell their lands. The state closed education and health public services in the localities and accused some local leaders of terrorism.

In this landscape of institutional weakness and lack of state legitimacy, local social organizations and networks became the main structures that mediate between the mining companies and the households. Moreover, because of the absence of other institutions with which to negotiate, both BHP and Rio Tinto preferred to negotiate with the *ronda campesina*. In contrast to the arguments of some writers (e.g., Damonte 2012), mining development does not necessarily lead to institutional weakening and fragmentation. To the contrary, in this case, and perhaps as an intended consequence, it has strengthened a key local organization.

Despite the ongoing transformations, *Granjinos* are becoming neither more individualistic nor more likely to enter into anonymous relations. For the *Granjinos*, kinship and other forms of social networks and membership (for instance, regional and local bonds and the membership of the *rondas campesina*) play a central role in their economic, social, and cultural lives. Even more, a sense of belonging and place attachment continues and the migrants in the city reproduce them through diverse cultural and sport activities, the maintenance of religious festivities, and the recreation of patterns of social differentiation.

Nevertheless, the transformations that mining development prompts do alter other aspects of the local social order. Gender relations are perhaps the most well known and important of those changes. New generations of women are increasingly challenging existing gender relations and roles in La Granja. This is particularly true of issues regarding economic autonomy and spatial mobility. There is a clash between traditional forms of social organization (namely, the patriarchal order, system of land inheritance and land ownership, and sexual division of labor) and imperatives of gender equality. Mining development—with the opening of employment and economic opportunities for women and the support for the creation of the women's patrols—has created new avenues for female autonomy. Of course, the new opportunities find resistance, from both men and women, and this sometimes leads to gender violence.

8.4 The limits of kinship

In the *Granjina* society, kinship plays a central role in the organization of its economic, social, and symbolic relations and in its spatial configuration. Certainly, the analyzed accounts show not only the importance of family networks for accessing and using land but also some of the complexities of the strategies. Land is not automatically inherited and the rights over it are not defined in a permanent way. People deploy strategies that allow a relatively fluid circulation of land among different family members, following specific circumstances and arrangements of need, position, and money, among other factors.

In the different gathered narratives, the use of family networks for the transmission and allocation of land is a key strategy. Nevertheless, there are two current processes that, to some extent, are in tension with kinship relations for defining the rules for space access. The first one is the coexistence of general formal norms (e.g., the national legal laws that regulate the transmission of property rights) with a parallel set of regulations dictated by the *ronda campesina*, norms that are not necessarily decided within the family units. Although the *ronda campesina* cannot oppose a family's decision to sell part of its land, it can, to some extent, regulate who is included in the list for negotiating with the mining company employment opportunities and other benefits. The second process is the increasing arrival of outsiders, such as schoolteachers, health workers, technicians, farmers from surrounding villages, and city fellows. These newcomers are not necessarily related to the local families and through marriage alliance seek to obtain full rights for accessing land and other benefits. Both processes could expand kinship boundaries but also mark the limits of family networks for the control of the space. The presence of the mining project, and the market forces that usually accompany it, fosters the expansion of both processes. We have examined some families that have combined kinship networks—that go beyond the inheritance system with their own kin—with market forms to control a significant number of properties in the context of mining development. Kinship fundamentally means alliance, an extended system of exchanging people, goods and services, which requires deference, kindnesses, relationships, and expectations in order to be maintained.

Kinship, however, is a regulatory system that governs over specific persons and not upon abstract individuals. Opposite to the principles of universal rights—which assume equal and general entitlements for all the citizens despite any consideration of race, age, faith, or gender—kinship requires the particularization of the individual into his/her personal attributes. A specific person is entitled to specific rights and duties owing to their position in a specific arrangement of sex, age, and kin. It is quite common, then, for kinship principles to be involved in conflict with universal categories, such as gender.

In La Granja, as in many other Andean societies, women are on the margins of the land ownership system (Bourque & Warren 1981; Deere & León de Leal 1981; Hamilton 1998). In contrast to men, women do not usually inherit or own land if they are single. They only do so once they get married and, consequently, their ability to access, control, and own land is limited to their civil status as spouses. In the 14 analyzed families, only two females accessed a property while single. Women were considered dependents of other men, first of their fathers and then of their husbands. Only after getting married would women leave their parents' household. The exceptions were single mothers or widowers, who had the chance to secure land rights on their own. Women were confined to the domestic sphere, where they oversaw domestic tasks and had to spin to produce clothes and blankets.

More recently, the villagers consider that women can buy land or properties directly from their owners, and the access to land leads to more possibilities for women to access economic autonomy, together with physical and political autonomy (United Nations 2010). However, it is worth mentioning that because they have fewer income opportunities, their capacity for saving is limited. Therefore, even though there are no legal restrictions for acquiring land and properties, structural and cultural barriers prevent them from accessing land in the same proportion to men. Nowadays, the opportunities for inhering land are also restricted. Two reasons could explain this situation. First, the enormous fragmentation of holdings makes it difficult to continue subdividing the land.[1] Second, farming activities are declining in importance, in comparison to nonfarming activities for the households' economy. In this sense, farming is increasingly becoming a female responsibility, in what is called the "feminization of the country" (Remy 2014), although men still control land ownership. And this ownership allows men to negotiate with the mining project.

Despite these difficulties, younger women are experiencing more freedom to own land without being engaged in a marital relationship. Market relations could imply liberating forces from the kinship yoke and the opening of spaces for the redefinition of gender relations and identities of the self. In doing this, nevertheless, they also unleash anxieties and fears that could result in friction and violence.

8.5 Hybrid identities and social positioning

The case study provides a good example of how ongoing social and economic transformations are helping to shape complex changes in collective and individual identities. In Peruvian society, groups and individuals tend to construct their social identities in relation to the position of other individuals and they do this within a hierarchical structure of labor, which values literacy and intellectual work over physical labor (Fuller 2001).[2] Racial considerations overlap with a system of classification and rate of labor and create a hierarchical world, which hegemonic perceptions see as a natural order (Nugent 2012). In this order, for instance, a company manager is "naturally" classified as *blanquilloso* (white) and is superior to a bus driver who is "naturally" classified as *zambo* (African descendant), who, in turn, is above a rural farmer or *campesino*, who is "naturally" described as *indio* (indigenous person from the Andes). However, unlike the apartheid system of the United States and South Africa—in which the "color line" marks two clearly defined and separate groups with its own distinctive social and phenotypical features—the racial class system in Peru is relational (Fuenzalida 1970b). A person is defined as white only in relation to another and he or she could be considered more indigenous in another context. For instance, a *Granjino* could be considered as *serrano* (indigenous) by a middle-class, white-collar employee of Cajamarca city, who in turn would be a *serrano* for a Limenian. In addition, any Limenian is classified as less white than a Western foreigner. Moreover, as Nugent suggests (2012), from the perspective of the elite, only their members are allowed to develop their life projects; to become what they want and fulfill their individual dreams. This perspective denies individuality to members of subaltern groups, who are condemned to reproduce their collective destiny. In this vision, the children of subaltern groups should reproduce the labor positions of their parents; for instance, as bus drivers or *campesinos* (Hopkins, van der Borght & Cavassa 1990).

Against this general background, it could be asked what happens when farming no longer defines the identities of many rural people, as is the case of the *Granjinos*. As has been said, land continues to be central to the economic life of the local families, not because of farming—this is to say, a labor relation—but because it allows the families to negotiate benefits with the mining company. In other words, many *Granjinos* are becoming less *campesinos* in relation to their position within the labor system. This raises the question of what kind of identities the *Granjinos* can construct if their socioeconomic position is no longer tied to productive relations.

The call for an ethnic identity is an option. Certainly, ethnicity crosses occupation and class distinctions and it has attracted diverse groups in Bolivia and Ecuador. This is especially true in current circumstances where the development of international and national legal frameworks provides incentives for ethnic formation, as in the case of the implementation of Convention 169 of the International Labour Organization (Castillo 2002). In the context of the rapid change unleashed by the development of the Tintaya mining project in Cusco, in the Peruvian southern Andes, Eduardo Cáceres (2014) finds a similar process of de-peasantization and the emergence of an ethnic discourse. An ethnic discourse, expressed in the construction of a K'ana identity (Cáceres 2013),[3] becomes part of the material for political struggles for the control of the local government and significant resources from the mining project among different groups.

A location-based identity is another option. Indeed, a sense of belonging to localized spaces has both shaped rural identities and contributed to the fragmentation that makes difficult the creation of broader alliances. In this sense, the *Granjinos* would develop new identities not as *campesinos* but as people linked to their place of origin. The creation of the Association of Granjinos in Chiclayo points in that direction. In addition to considerations of place—such as, for instance, those born in the village of La Granja and those who have migrated from other rural areas—the *Granjinos* use occupation criteria, such as *empresario* or entrepreneur, to identify themselves and to distinguish themselves from others.

La Granja's case suggests that mining development can trigger struggles over personal positioning in a hierarchical system where the ethnic component still plays an important role. However, the appeal to ethnic considerations is not an automatic and necessary strategy. The *Granjinos* have developed different strategies, such as positioning themselves in relation to other members of the locality—for instance, opposing "traditional landowners" versus newcomers—and vis-à-vis the company—playing the role of "strategic marginality" (Mallon 1996). The *Granjinos* do not have one static identity but use different ones in a fluid manner according to the interlocutor and the circumstances in what could be called "strategic multiplicity", using Charles Hale's (1997) concept.

In brief, instead of using essential and fixed categories, local people act following relational categories and navigate within hybrid systems to define themselves; they perform the game of "identity politics" (Hale 1997). In any case, the *Granjino* families do not accept being labeled as *campesinos* in a derogative manner. The statement that a social worker from Cambior made to a female villager was not simply about the urban comfort but about changing their identities from *campesinos* to people "owning a place that you deserve". Through their actions and strategies, the *Granjinos* are defying their destiny and are constructing collective and individual life projects. As shown in the case of La Granja, mining development can prompt substantial transformations of the rural landscape, fostering an increase of spatial mobility and, more significant, social mobility. In a hierarchical social environment where acknowledgment of equal and universal citizenship is weak and resisted, people take on different identities to climb up the social ladder.

8.6 Concluding observations and research agenda

Our comprehension of rural transformations in the context of mining development emphasizes three features: space, time, and a view from the bottom.

Spatial dimensions are particularly relevant for the understanding of the contradictory forms of development that the fixidity/fluidity of mineral extraction generates (Macintyre 2018). The access and appropriation of land, the movement of people, goods, and ideas, the formation of productive spaces, and the construction of place-based Arcadias and utopias are at the center of many local struggles. Integrating geographical and historical aspects into the same analysis enables a rich understanding of some of the contradictions of capital in its ceaseless spatial and temporal movement. It allows the link between global and structural processes (such as commodity consumption, cycles of mining expansion, and implementation of free-trade policies) with collective and personal experiences (including displacement, social mobility, and sense marginality and isolation in urban settings) and strategies (for instance, land subdivision, use of kinship, temporary migration, and double residence). The reading through the lens of the theory of the "production of space" facilitates connecting social actions with territory in a fluctuating movement, what Arturo Escobar (2011) calls "sitting culture in places".

This reading calls for the return of regional studies—such as the formation of land and labor markets or the establishment of broad political alliances—for the examination of mineral extraction and the responses that local families deploy.

A view from the bottom favors the analysis of the multiple strategies that nuclear families use over time. To be sure, even though many anthropologists have stressed the importance of kinship and social networks in their examination of Andean societies, much current examination of the social effects of mining development has adopted a standard approach of social impact assessment (SIA). Our focus on nuclear families and their social networks offers a variety of advantages. First, typically, SIA assigns positive or negative attributes to the impacts from an external perspective when changes are generally complex and ambivalent and depend on particular circumstances of the individuals and families (for instance, migratory experiences). Second, although conventional SIAs investigate impacts over territories, following environmental frameworks of direct and indirect areas of influence, social changes affect people and networks rather than contiguous geographical units. People's mobility—and their use of double-residence strategies—has substantial implications for social and economic development of families and regions (Castillo & Brereton 2018a) and is an undersearched theme. The close examination of family networks is a technique especially fruitful in cases that exhibit high spatial mobility and informal activities that are difficult to capture through static census.

In addition, cross-temporal studies that follow local families in the context of mining development are extremely useful. The combination and complementarity of qualitative and quantitative studies is valuable. For instance, quantitative analysis could measure average variations in land size and land value among households over time or determine the fluctuations and distribution of household income, not only in absolute terms but also in the relative weight of farming and nonfarming activities. Qualitative studies could examine the household strategies involved on those changes, including the insertion of women into the labor market or the local institutional constrains for the formation of land markets.

Third, the use of nuclear family as the unit of analysis allows the examination of the interrelations between individual and collective forces that shape social and economic decisions. Fourth, the focus on family networks facilitates observation of the intersection of different layers—of power, gender, symbolic and economic relations—with agents that operate at different spatial scales. Certainly, a detailed assessment of the implications for gender relations of the transformations that mining development fosters in each major area of the local social life should appear in the research agenda: for instance, the political opportunities and challenges that women face with the arrival of mining projects in their localities, the changes and continuities of gender roles in productive and reproductive labor, or the gender violence that mining development could exacerbate among local families and within households.

A final advantage of our focus is that it departs from a company-centered and top-down approach to emphasize local perspectives and highlight local actors' agency. Our analysis demonstrates that social and spatial transformations prompted by mining development in rural societies cannot be merely understood as the mechanical result of the effect of global forces. People's resourcefulness and bottom-up processes are crucial for comprehending the outcomes. In the La Granja case, kinship and local networks have acted as a safety net and a distribution system of economic benefits. In consequence, with the use of extended kinship networks, the local families—and not exclusively state agencies and company policies— have been able to access a significant portion of the distribution of goods and services, as well as influencing some of the decisions dealing with the development of the project at local scale.

In line with what Massey (1994) argues in her study of Sao Paulo's *favelas*, the families of La Granja are contributing to construct a global sense of place. Their actions could be physically limited to a regional territory; however, these actions are performed vis-à-vis actions that have repercussions at a global level. Rio Tinto's final decision to go ahead or not with the plans for exploiting one of the biggest copper ore deposits in the world would be closely linked to China's mineral appetite; but the *Granjinos* are the ones that experience and make sense of those global decisions on the ground.

In 2014, Rio Tinto's managers in their London headquarters took the decision to put the mining project on hold; a decision that could be attributed to financial constraints of the corporation or the global downturn in mineral prices. However, for the local families of La Granja, as many other ones around the world, this cycle would be just a chapter in their longer history of spatial transformation, which they experience and actively shape.

Notes

1 In some of the cases examined, a couple registers the new properties bought in the name of their children, but only the male ones.

2 The access to formal education in Spanish—the hegemonic language in the country over native languages—and the mastering of literacy—in contrast to oral forms—are central components in the construction of the power structure of Peruvian social exclusion. The school system endlessly recounts and reinforces the story of the last Inca, Atahualpa, in his encounter with the Spanish conqueror Francisco Pizarro in the moment before he was captured in Cajamarca. The mythology indicates that Atahualpa threw aside the Bible preached by a Catholic priest because he was not able to hear the word of God from the book. It is the foundational image in the discourse of the superiority of the Spanish and literate world over the indigenous and oral one (Nugent 2010). It is not for nothing that after the fight for land, access to education has been the most important struggle that Andean *campesinos* in Peru have embraced (Degregori 1986).

3 The ethnic group than formed the Pre-Hispanic chiefdoms of Canas and Canches in the region.

Appendix

Table A.1 Interviewed families

Number	Name	Type	Location	Brief profile
1	Resistant_1	Resistant	La Granja	Young family. The male informant is from La Granja and did not sell the land to Cambior.
2	Resistant_2	Resistant	La Granja	Middle-aged family. Both informants are from La Granja and did not sell the land to Cambior.
3	Resistant_3	Resistant	La Granja	Mature family. The female informant is from La Granja and did not sell the land to Cambior.
4	Returnee_1	Returnee	La Granja	Middle-aged family. The male informant is from La Granja. They sold their land to Cambior and returned to the locality at the time of BHP Billiton.
5	Returnee_2	Returnee	La Granja/ La Lima	Mature family. Both family heads are from La Granja, in the La Lima sector. They sold their land to Cambior and returned at the time of BHP Billiton.

Number	Name	Type	Location	Brief profile
6	Opportunistic_1	Opportunistic	La Granja	Mature family. They are from neighboring localities and arrived in La Granja at the time of BHP Billiton.
7	Opportunistic_2_LG	Opportunistic	La Granja	Young family. The male informant is from Chota but moved to La Granja and married a woman of the area with the arrival of Rio Tinto.
8	Migrant_1	Migrant	Querocoto	Mature family. They sold their land to Cambior and established in Querocoto.
9	Migrant_2	Migrant	Chiclayo	Mature family. They migrated to Chiclayo before Cambior. The male informant trades between Chiclayo and La Granja.
10	Migrant_3	Migrant	Chiclayo	Young family. The female informant is from Pariamarca and her partner is from La Granja. She lives and works in Chiclayo, where she takes care of her baby.
11	Migrant_4	Migrant	Ojo de Toro	Middle-aged family from Paraguay. They moved to Ojo de Toro, a rural costal area close to Chiclayo in 2000.
12	Regional_1	Regional	Querocoto	Mature family from Querocoto. They own a small hardware store.
13	Regional_2	Regional	Querocoto	Middle-aged family from Querocoto. They own a mixed business of hardware, convenience store, and bakery.
14	Regional_3	Regional	Querocoto	Young family from Querocoto. They own a small convenience store.

References

Alber, E 1999, *¿Migración o movilidad en Huayopampa?: nuevos temas y tendencias en la discusión sobre la comunidad campesina en los Andes*, IEP, Lima.

Alberti, G & Mayer, E (comps) 1974, *Reciprocidad e intercambio en los Andes peruanos*, IEP, Lima.

Albrecht, G 2005, 'Solastalgia, a new concept in human health and identity', *Philosophy Activism Nature*, vol. 3, pp. 41–44.

Allen, J, Massey, DB & Cochrane, A 1998, *Rethinking the region*, Routledge, New York.

Altamirano, T 1977, *Estructuras regionales, migración y asociaciones regionales en Lima*, PUCP, Lima.

Altamirano, T 2000, *Migration, remittances and development in times of crisis*, UNFPA/CISEPA/PUCP, Lima.

Altamirano, T 2010, *Liderazgo y organizaciones de peruanos en el exterior: culturas transnacionales e imaginarios sobre el desarrollo*, PromPerú/PUCP, Lima.

Arana, M 2011, 'Como nos llevaron al Conga no va y los nuevos caminos que debemos andar', viewed 22 January 2012, http://lamula.pe/2011/12/04/como-nos-llevaron-al-conga-no-va-y-los-nuevos-caminos-que-debemos-andar/lamula

Arellano-Yanguas, J 2011, *¿Minería sin fronteras?: conflicto y desarrollo en regiones mineras del Perú*, UARM/IEP/PUCP, Lima.

Assadourian, CS 1982, *El sistema de la economía colonial: mercado interno, regiones y espacio económico*, IEP, Lima.

Bainton, N 2010, *The Lihir destiny: cultural responses to mining in Melanesia*, ANU E Press, viewed 24 January 2013, http://epress.anu.edu.au/lihir_destiny_citation.html

Bainton, NA & Banks, G 2018, 'Land and access: a framework for analysing mining, migration and development in Melanesia', *Sustainable Development*, vol. 26, no. 5, pp. 450–60, doi: 10.1002/sd.1890

Ballard, C & Banks, G 2003, 'Resource wars: the anthropology of mining', *Annual Review of Anthropology*, vol. 32, pp. 287–317.

Bebbington, A (ed) 2007, *Minería, movimientos sociales y respuestas campesinas: una ecología política de transformaciones territoriales*, IEP/CEPES, Lima.

Bebbington, A 2009, 'The new extraction: rewriting the political ecology of the Andes?', *NACLA Report on the Americas*, vol. 42, no. 5, pp. 12–40.

Bebbington, A & Bury, J 2013a, 'Political ecologies of the subsoil' in Bebbington, A & Bury, J (eds), *Subterranean struggles: new dynamics of mining, oil, and gas in Latin America*, pp. 1–25, The University of Texas Press, Austin.

Bebbington, A & Bury, J (eds) 2013b, *Subterranean struggles: new dynamics of mining, oil, and gas in Latin America*, The University of Texas Press, Austin.

Bebbington, A & Humphreys Bebbington, D 2011, 'An Andean avatar: post-neoliberal and neoliberal strategies for securing the unobtainable', *New Political Economy*, vol. 16, no. 1, pp. 131–45.

Bebbington, A & Humphreys Bebbington, D 2018, 'Mining, movements and sustainable development: concepts for a framework', *Sustainable Development*, vol. 26, no. 5, pp. 441–49, doi: 10.1002/sd.1888

Bebbington, A & Williams, M 2008, 'Water and mining conflicts in Peru', *Mountain Research and Development*, vol. 28, no. 3–4, pp. 190–95.

Berg, UD & Paerregaard, K (eds) 2005, *El 5to Suyo: transnacionalidad y formaciones diaspóricas en la migración peruana*, IEP, Lima.

Berman, M 1982, *All that is solid melts into air: the experience of modernity*, Simon & Schuster, New York.

Boege, V, Brown, A, Clements, K & Nola, A 2008, 'On hybrid political orders and emerging states: state formation in the context of 'fragility'', Berghof Research Center for Constructive Conflict Management, viewed 23 August 2012, www.berghof-handbook.net/ documents/publications/boege_etal_handbook.pdf

Boege, V & Franks, DM 2012, 'Reopening and developing mines in post-conflict settings: the challenges of company-communities relations' in Lujala P & Rustad SA (eds), *High-value natural resources and post-conflict peacebuilding*, pp. 87–120, Earthscan, Abingdon.

Bonilla, H 1974, *El minero en los Andes: una aproximación a su estudio*, IEP, Lima.

Bourque, SC & Warren, KB 1981, *Women of the Andes: patriarchy and social change in two Peruvian towns*, Women and Culture Series, University of Michigan Press, Ann Arbor.

Brain, KA 2017, 'The impacts of mining on livelihoods in the Andes: a critical overview', *The Extractive Industries and Society*, vol. 4, pp. 410–18, doi: 10.1016/j.exis.2017.03.001

Bridge, G 2004, 'Contested terrain: mining and the environment', *Annual Review of Environment and Resources*, vol. 29, pp. 205–59.

Brundenius, C 1972, 'The anatomy of imperialism: the case of the multinational mining corporations in Peru', *Journal of Peace Research* vol. 9, no. 3, pp. 189–207.

Burneo, ML & Chaparro, A 2011, *Michiquillay: dynamics of transference and changes in land use and valuation in the context of mining expansion in an Andean campesino community*, CISEPA/CEPES/CIRAD/International Land Coalition, viewed 16 November 2012, www.landcoalition.org/sites/default/files/publication/1023/MICHIQUILLAY_ENG_web_11.03.11.pdf

Bury, J 2004, 'Livelihoods in transition: transnational gold mining operations and local changes in Cajamarca, Peru', *The Geographical Journal*, vol. 170, no. 1, pp. 78–91.

Bury, J 2011, 'Minería, migración y transformaciones en los medios de subsistencia en Cajamarca, Perú', in Bebbington, A (ed), *Minería, movimientos sociales y respuestas campesinas: una ecología política de transformaciones territoriales*, pp. 261–307, 2nd edn, IEP, Lima.

Bury, J & Kolff, A 2002, 'Livelihoods, mining and peasant protests in the Peruvian Andes', *Journal of Latin American Geography*, vol. 1, no. 1, pp. 3–16.

Cáceres, E, 2013, 'De corredor minero a proyecto regional: Espinar y las provincias altas del Cusco' in Cáceres, E & Rojas, J (eds), *Minería, desarrollo y gestión municipal en Espinar*, pp. 15–93, OXFAM/SER, Lima.

Cáceres, E, 2014, 'Territorio, historia y conflicto minero: Hualgayoc y Espinar', viewed 5 September 2014, www.noticiasser.pe/20/08/2014/territorio-historia-y-conflicto-minero-hualgayoc-y-espinar

Calderón, J 2005, *La ciudad ilegal: Lima en el siglo XX*, Fondo Editorial de la Facultad de Ciencias Sociales UNMSM, Lima.

Castilla, LM 2012, 'Importancia de la minería en el desarrollo peruano', Paper presented to the 10th Peru: International Gold Symposium, Lima, 14–16 May, viewed 15 February 2013, www.mef.gob.pe/contenidos/comun_notp/presentaci/2012/simposium_oro.pdf

Castillo, G 2002, 'International lending institutions, the State, and indigenous movements in the Latin America of the 1990s', Paper based on the research conducted under the supervision of Karl Offen, The University of Oklahoma, viewed 11 July 2014, www.academia.edu/7355412/International_Lending_Institutions_the_State_and_Indigenous_Movements_in_the_Latin_America_of_the_1990s

Castillo, G 2006, 'Se vende oro: la creación de espacios contestados en la promoción de la minería peruana', in Cánepa, G & Ulfe, ME (eds), *Mirando la esfera pública desde la cultura en el Perú*, pp. 95–106, CONCYTEC, Lima.

Castillo, G 2015, *Transforming andean space: local experiences of mining development in Peru*, PhD Thesis, Sustainable Minerals Institute, The University of Queensland, Brisbane, doi: 10.14264/uql.2015.958

Castillo, G & Brereton, D 2018a, 'Large-scale mining, spatial mobility, place-making and development in the Peruvian Andes', *Sustainable Development*, vol. 26, no. 5, pp. 501–5, doi: 10.1002/sd.1895

Castillo, G & Brereton, D 2018b, 'The country and the city: mobility dynamics in mining regions', *Extractive Industries and Society*, vol. 5, no. 2, pp. 307–16, doi: 10.1016/j.exis.2018.02.009

122   *References*

Castillo, G & Soria, L 2011, *Report: gender justice in consultation processes for extractives industries in Bolivia, Ecuador and Peru*, Oxfam, Lima.

Cetraro, J, Castro, E & Chávez, J (eds) 2007, *Nueva ruralidad y competitividad territorial*, Ideas, Lima.

Clastres, H 1989, *La tierra sin mal: el profetismo tupí-guaraní*, Ediciones del Sol/Ediciones de Aquí a la Vuelta, Buenos Aires.

Clifford, J 1994, 'Diasporas', *Cultural Anthropology*, vol. 9, no. 3, pp. 302–38.

Comisión de la Verdad y la Reconciliación 2004, *Hatun Willakuy: versión abreviada del Informe final de la Comisión de la Verdad y Reconciliación, Perú*, CVR, Lima.

Comisión Económica para América Latina y el Caribe (CEPAL) 2014, *Informe anual 2013–2014: el enfrentamiento de la violencia contra las mujeres en América Latina y el Caribe*, CEPAL, Santiago de Chile.

Contreras, C 1988, *Mineros y campesinos en los Andes: mercado laboral y economía campesina en la sierra central. Siglo XIX*, IEP, Lima.

Contreras, C 1995, *Los mineros y el rey: los Andes del norte, Hualgayoc 1770–1825*, IEP, Lima.

Cook, ND 2010, *La catástrofe demográfica andina: Perú 1520–1620*, PUCP, Lima.

Coronil, F 1997, *The magical state: nature, money, and modernity in Venezuela*, The University of Chicago Press, Chicago.

Cortes, G 2004, *Partir para quedarse: supervivencia y cambio en las sociedades campesinas andinas de Bolivia*, Institut Français d'Études Andines/Institut de Recherche pour le Développement/Plural Editores, La Paz, doi: 10.4000/books.ifea.4368

Cosgrove, D 1985, *Social formation and symbolic landscape*, Barnes & Noble Books, Totowa.

Cresswell, T 2004, *Place: a short introduction*, Blackwell, Oxford.

Damonte, G 2007, 'Minería y política: la recreación de luchas campesinas en dos comunidades andinas', in Bebbington, A (ed), *Minería, movimientos sociales y respuestas campesinas: una ecología política de transformaciones territoriales*, pp. 117–62, IEP, Lima.

Damonte, G 2008, 'Industrias extractivas: el caso de la gran minería. Ponencia balance', in Damonte, G, Gómez, R & Fulcram, B (eds), *Perú: el problema agrario en debate, SEPIA XII*, pp. 19–77, SEPIA, Lima.

Damonte, G 2012, 'Dinámicas rentistas: transformaciones institucionales en contextos de proyectos de gran minería', GRADE, viewed 11 July 2014, http://bibliotecavirtual.clacso.org.ar/Peru/grade/20121109040224/30_damonte.pdf

Damonte, G & Castillo, G 2011, 'Presentación: una mirada antropológica a las industrias extractivas en los Andes', *Anthropologica*, vol. 28, no. 28, pp. 5–19.

De Echave, J, Diez, A, Huber, L, Revesz, B, Lanata, XR & Tanaka, M 2009, *Minería y conflicto social*, CBC/CIPCA/CIES/IEP, Lima.

De la Cadena, M 1991, 'Las mujeres son más indias: etnicidad y género en una comunidad del Cusco', *Revista Andina*, vol. 9, no. 1, pp. 7–29.

Deere, CD 1990, *Household and class relations: peasant and landlords in northern Peru*, University of California Press, Berkeley.

Deere, CD & León de Leal, M 1981, 'Peasant production, proletarization, and the sexual division of labor in the Andes', *Signs*, vol. 7, no. 2, pp. 338–60.

Degregori, CI 1986, 'Del mito de Inkarri al mito del progreso: poblaciones andinas, cultura e identidad nacional', *Socialismo y Participación*, no. 36, pp. 49–56.

Degregori, CI, Lynch, N & Blondet, C 1986, *Conquistadores de un nuevo mundo: de invasores a ciudadanos en San Martín de Porres*, IEP, Lima.

Deustua, J 1994, 'Mining markets, peasants, and power in Nineteenth-Century Peru', *Latin American Research Review*, vol. 29, no.1, pp. 29–53.

DeWind, J 1987, *Peasants become miners: the evolution of industrial mining systems in Peru*, Garland Publishing, New York/London.

Dollfus, O 1981, *El reto del espacio andino*, IEP, Lima.

Eguren, F 2006, *La reforma agraria en el Perú*, Consulta de expertos en reforma agraria en América Latina, 11 y 12 de diciembre, Oficina Regional de la FAO para América Latina y el Caribe, Santiago de Chile.

Escobal, J 2001, 'The determinants of nonfarm income diversification in rural Peru,' *World Development*, vol. 29, no. 3, pp. 497–508.

Escobar, A 1995, *Encountering development: the making and unmaking of the Third World*, Princeton University Press, Princeton.

Escobar, A 2011, 'Culture sits in places: reflections on globalism and subaltern strategies of localization', *Political Geography*, vol. 20, pp. 139–74.

Farrell, L, Sampat, P, Sarin, R & Slack, K 2004, *Dirty metals: mining, communities and the environment*, Earthworks/Oxfam America, Washington/Boston.

Ferguson, J 1990, *The anti-politics machine: "development", depoliticization, and bureaucratic power in Lesotho*, Cambridge University Press, Cambridge.

Ferguson, J 1999, *Expectations of modernity: myths and meanings of urban life on the Zambian Copperbelt*, University of California Press, Berkeley, CA.

Fernandez, JP 2019, 'Respuesta de Rio Tinto sobre La Granja: "Hemos ejercido la opción de prórroga semestral"', *Energiminas*, viewed 22 September 2019, www.energiminas.com/respuesta-de-rio-tinto-sobre-la-granja-hemos-ejercido-la-opcion-de-prorroga-semestral/

Fisher, JR 1971, 'La rebelión de Túpac Amaru y el programa de la reforma imperial de Carlos III', *Anuario de Estudios Americanos*, vol. 28, pp. 405–21.

Flores Galindo, A 1977, *Arequipa y el sur andino: ensayo de historia regional (Siglos XVIII-XX)*, Horizonte, Lima.

Flores Galindo, A 1993 [1974], 'Los mineros de Cerro de Pasco, 1900–1930: un intento de caracterización social', in Rivera, C & Martínez, M (eds), *Alberto Flores Galindo: obras completas*, pp. 1–229, vol. I, Fundación Andina/ SUR Casa de Estudios del Socialismo, Lima.

Franks, D 2015, *Mountain movers: mining, sustainability and the agents of change*, Routledge, Abingdon.

Fuenzalida, F 1970a, 'La matriz colonial de la comunidad de indígenas peruana: una hipótesis de trabajo', *Revista del Museo Nacional*, vol. 35, pp. 92–123.

Fuenzalida, F 1970b, 'Poder, raza y etnia en el Perú contemporáneo', in Fuenzalida, F, Mayer, E, Escobar, G, Bourricaud, F & Matos Mar, J (eds), *El indio y el poder en el Perú*, pp. 15–87, Perú Problema 4, IEP/Moncloa-Campodónico, Lima.

Fuenzalida, F, Valiente, T, Villarán, JL, Golte, J, Degregori, CI & Casaverde, J 1982, *El desafío de Huayopampa: comuneros y empresarios*, IEP, Lima.

Fuertes, P & Velazco, J 2013, 'Les menages au Pérou (1999–2009): transformations en période d´expansion économique', *Problémes d´ Amérique Latine. Pérou: émergence économique et zones d´ombre*, vol. 88, pp. 25–53.

Fuller, N 2001, *Masculinidades: cambios y permanencias. Varones de Cuzco, Iquitos y Lima*, PUCP, Lima.

García, A 2007, 'El síndrome del perro del hortelano', *El Comercio*, 28 October, viewed 1 October 2014, www.justiciaviva.org.pe/userfiles/26539211-Alan-Garcia-Perez-y-el-perro-del-hortelano.pdf

Giarraca, N (coord) 2002, *¿Una nueva ruralidad en América Latina?*, CLACSO, Buenos Aires.

Gil, V 2009, *Aterrizaje minero*, IEP, Lima.

Glave, M 2008, "Valor y renta de la tierra en los Andes peruanos: reflexiones en torno a la nueva minería", in Damonte, G, Fulcrand, B & Gómez, R (eds), *Perú: el problema agrario en debate - SEPIA XII*, pp. 182–201, SEPIA, Lima.

Godoy, R 1995, 'Mining: anthropological perspectives', *Annual Review of Anthropology*, vol. 14, pp. 199–217.

Golte, J & Adams, N 1987, *Los caballos de Troya de los invasores: estrategias campesinas en la conquista de la gran Lima*, IEP, Lima.

Grammont de, HC 2004, 'La nueva ruralidad en América Latina', *Revista Mexicana de Sociología*, vol. 66, Special Number, pp. 279–300.

Guevara, J n.d., *Añoranzas de mi tierra*, Manuscript.

Gustafson, B & Guzmán Solano, N 2018, 'Mining movements and political horizons in the Andes: articulation, democratisation, and worlds otherwise', in Deonandan, K & Dougherty, ML (eds), *Mining in Latin America: critical approaches to the new extraction*, pp. 141–59, Routledge, Abingdon.

Hale, C 1997, 'Cultural politics of identity in Latin America', *Annual Review of Anthropology*, vol. 26, pp. 567–90.

Hamilton, S 1998, *The two-headed household: gender and rural development in the Ecuadorean Andes*, University of Pittsburgh Press, Pittsburgh, PA.

Harvey, D 1985, *Consciousness and the urban experience: studies in the history and theory of capitalist urbanization*, Johns Hopkins University Press, Baltimore, MD.

Harvey, D 1999, *The limits to capital*, Verso, London.

Helfgott, F 2013, *Transformations in labor, land and community: mining and society in Pasco, Peru, 20th century to the present*, PhD thesis, University of Michigan, Ann Harbor, MI.

Himley, M 2010, *Frontiers of capital: mining, mobilization, and resource governance in Andean Peru*, PhD Dissertation, Syracuse University, Syracuse, NY.

Himley, M 2014, 'Los límites de la solución tecnológica: minería, agua y poder en el Perú', in Perreault, T (ed), *Minería, agua y justicia social en los Andes: experiencias comparativas de Perú y Bolivia*, pp. 59–79, Justicia Hídrica-Paraguas/Centro de Estudios Regionales Andinos Bartolomé de las Casas, Cusco.

Hirsch, E 2017, 'Remapping the vertical archipelago: mobility, migration, and the everyday labor of Andean development', *Journal of Latin American and Caribbean Anthropology*, vol. 23, no. 1, pp. 189–208.

Hobsbawn, E 1974, 'Peasant land occupations', *Past and Present*, vol. 62, pp. 120–52, viewed 15 October 2012, http://past.oxfordjournals.org

Hopkins, R, van der Borght, D & Cavassa, A 1990, *La opinión de los campesinos sobre le política agraria: problemas, alternativas y rol de la organización*, Documento de Trabajo no. 37, Serie Economía no. 12, IEP, Lima.

Huerta-Mercado, A 2006, 'Ciudad abierta: lo popular en la ciudad peruana', in DESCO, *Perú hoy: las ciudades en el Perú*, pp. 129–53, DESCO, Lima.

Hurtado, I 2000, 'Dinámicas territoriales: afirmación de las ciudades intermedias y surgimiento de los espacios locales', in Brack, A, Hurtado, I & Trivelli, C (eds), *Perú: el problema agrario en debate, SEPIA VIII*, pp. 19–67, Seminario Permanente de Investigación Agraria, ITDG, Lima.

Instituto Nacional de Estadística e Informática 1993, *Censos nacionales 2007: IX de población y IV de vivienda*, Data file, viewed 26 November 2014, www. inei.gob.pe/estadisticas/censos/

Instituto Nacional de Estadística e Informática 2007, *Censos nacionales 2007: XI de población y VI de vivienda*, Data file, viewed 26 November 2014, www. inei.gob.pe/estadisticas/censos/

Instituto Nacional de Estadística e Informática 2012a, *IV Censo Nacional Agropecuario 2012*, Data file, viewed 16 October 2014, http://censos.inei. gob.pe/Cenagro/redatam/

Instituto Nacional de Estadística e Informática 2012b, *Perú: encuesta nacional de hogares sobre condiciones de vida y pobreza 2012*, Data file, viewed 16 October 2014, http://webinei.inei.gob.pe/anda_inei/index.php/catalog/195

Jacobsen, N 1993, *Mirages of transition: the Peruvian altiplano 1780–1930*, The University of California/Hants, Berkeley/Ashgate.

Kervyn, B 1996, 'La economía campesina en los Andes peruanos: teorías y políticas', in Morlon, P (comp), *Comprender la agricultura campesina en los Andes Centrales Perú – Bolivia*, pp. 424–56, Institut Français d´Études Andines/Centro de Estudios Regionales Andinos Bartolomé de las Casas, Cusco.

Kervyn, B & Equipo del CEDEP AYLLU 1989, 'Campesinos y acción colectiva: la organización del espacio en comunidades de la sierra sur del Perú', *Revista Andina*, vol. 7, no. 1, pp. 7–81.

Kirshenblatt-Gimblett, B 1994, Spaces of dispersal, *Cultural Anthropology*, vol. 9, pp. 339–44.

Kuramoto, J 1999, *Las aglomeraciones productivas alrededor de la minería: el caso de Minera Yanacocha S.A.*, GRADE, Lima.

Lefebvre, H 1991, *The production of space*, Blackwell Publishing, Malden.

Lefebvre, H 2002 [1968], *Everyday life in the modern world*, translated by Rabinovitch, S, with an introduction by Wander, P, Continuum, London.

Li, F 2015, *Unearthing conflict: corporate mining, activism, and expertise in Peru*, Duke University Press, Durham, NC.

Long, N & Roberts, B 1984, *Miners, peasants and entrepreneurs: regional development in the Central Highlands of Peru*, Cambridge University Press, Cambridge.

Lovell, AM 2010, 'Diaspora', in Warf, B (ed), *Encyclopedia of geography*, pp. 735–36, SAGE, Thousand Oaks, CA.

Macintyre, M 2018, 'Afterwords. Places, migration and sustainability: anthropological reflections on mining and moving', *Sustainable Development*, vol. 26, no. 5, pp. 501–05, doi: 10.1002/sd.1895

Mallon, FE 1983, *The defense of community in Peru's central highlands: peasant struggle and capitalist transition, 1860–1940*, Princeton University Press, Princeton, PA.

Mallon, FE 1996, 'Constructing *mestizaje* in Latin America: authenticity, marginality and gender in the claiming of ethnic identities', *Journal of Latin American Anthropology*, vol. 2, no. 1, pp. 170–81.

Manrique, N 1987, *Mercado interno y región: la sierra central 1820–1930*, DESCO, Lima.

Marcus, G 1995, 'Ethnography in/of the world system: the emergence of multi-sited ethnography', *Annual Review of Anthropology*, vol. 24, pp. 95–117.

Massey, DB 1994, *Space, place and gender*, Polity Press, Cambridge.

Masuda, S, Shimada, I & Morris, C (eds) 1985, *Andean ecology and civilization: an interdisciplinary perspective on Andean ecological complementarity*, University of Tokyo Press, Tokyo.

Matos Mar, J & Mejía, JM 1984, *Reforma agraria: logros y contradicciones, 1969–1979*, 2nd edn, IEP, Lima.

Mayer, E 1996, 'Zonas de producción: autonomía individual y control comunal', in Morlon, P (comp), *Comprender la agricultura campesina en los Andes Centrales: Perú – Bolivia*, pp. 154–70, Institut Français d'Études Andines/ Centro de Estudios Regionales Andinos Bartolomé de las Casas, Cusco.

Mayer, E 2002, *The articulated peasant: household economies in the Andes*, Westview Press, Cambridge.

Mayer, E & Bolton, R (eds) 1980, *Parentesco y matrimonio en los Andes*, PUCP, Lima.

McLean, SL, Schultz, DA & Steger, MB (eds) 2002, *Social capital: critical perspectives on community and "Bowling alone"*, New York University Press, New York.

McMahon, G & Remy, F (ed) 2001, *Large mines and the community: socioeconomical and environmental effects in Latin America, Canada, and Spain*, The World Bank, Washington, D.C.

Mendoza, A 2011, *Discussing local content: lessons drawn and recommendations based on experiences in Peru and worldwide*, Draft Report, RWI, Lima.

Mitchell, K 1997, 'Different diasporas and the hype of hybridity', *Environment and Planning D: Society and Space*, vol. 15, pp. 533–53.

Moore, DS 1998, 'Subaltern struggles and the politics of place: remapping resistance in Zimbabwe's Eastern highlands', *Cultural Anthropology*, vol. 13,
no. 3, pp. 344–81.

Murra, J 1975, *Formaciones económicas y políticas del mundo andino*, IEP, Lima.

Murra, J 2002, *El mundo andino: población, medio ambiente y economía*, IEP/
PUCP, Lima.

Nugent, G 2010, *El orden tutelar: sobre las formas de autoridad en América
Latina*, CLACSO/DESCO, Lima.

Nugent, G 2012, *El laberinto de la choledad: páginas para entender la desigual-
dad*, 2nd edn, UPC, Lima.

Offen, K 2003, 'Narrating place and identity, or mapping Miskitu land claims
in Northeastern Nicaragua', *Human Organization*, vol. 62, pp. 382–92.

Offen, K 2004, 'Historical political ecology: an introduction', *Historical
Geography*, vol. 32, pp. 19–42.

O'Phelan, S 2012, *Un siglo de rebeliones anticoloniales: Perú y Bolivia,
1700–1783*, Institute Français d'Études Andines, Lima.

Ossio, J 1981, 'Parentesco y matrimonio en los Andes', *Allpanchis*, vol. 17–18,
pp. 235–43.

Pasco-Font, A, Diez, A, Damonte, G, Fort, R & Salas, G 2001, 'Peru: learn-
ing by doing', in McMahon, G & Remy, F (eds), *Large mines and the community socioeconomic and environmental effects in Latin America, Canada
and Spain*, pp. 143–97, IDRC/World Bank, Ottawa.

Peet, R 1996, 'A sign taken for history: Daniel Shays' Memorial in Petersham,
Massachusetts', *Annals of the Association of American Geographers*, vol. 86,
no. 4, pp. 21–43.

Pintado, MA 2014, 'Diversificación de ingresos en los hogares agropecuarios:
el agro perdió peso', *La Revista Agraria*, vol. 11, no. 165, pp. 3–4.

Remy, MI 2014, '¿Feminización de la agricultura peruana?', *La Revista
Agraria*, vol. 14, no. 158, pp. 8–9.

Rénique, JL 2004, *La batalla por Puno: conflicto agrario y nación en los Andes
peruanos 1866–1995*, IEP/SUR/CEPES, Lima.

Rio Tinto 2009, *Why gender matters: a resource guide for integrating gender
considerations into communities work at Rio Tinto*, Rio Tinto, Melbourne.

Rivers, WHR 1968, *Kinship and social organization*, The Athlone Press, London.

Roberts, B 1978, *City of peasants: the political economy of Third World urbanization*, Edward Arnold, London.

Rodríguez, A, Riofrío, G & Welsh, E 1973, *De invasores a invadidos*, DESCO,
Lima.

Rosser, A 2006, *The political economy of the resource curse: a literature survey*,
Working Paper no. 268, IDS, Sussex.

Rumsey, A & Weiner, J (eds) 2004, *Mining and indigenous lifeworlds in Australia and Papua New Guinea*, Sean Kingston Publishing, Wantage.

Salas, G 2008, *Minería y dinámica social*, IEP, Lima.

Salas, G 2010, 'La embriaguez del canon minero: La política distrital en San
Marcos a doce años de la presencia de Antamina', *Anthropologica*, vol. 28,
no. 28, pp. 111–38.

Salas, M 1998, *Estructura colonial del poder español en el Perú: Huamanga (Ayacucho) a través de sus obrajes. Siglos XVI-XVIII*, PUCP, Lima.

Sandoval, P 2000, 'Los rostros cambiantes de la ciudad: cultura urbana y antropología en el Perú', in Degregori, CI (ed), *No hay país más diverso: compendio de antropología peruana*, pp. 278–329, Red para el Desarrollo de las Ciencias Sociales, Lima.

Schorr, B & Dietz, K 2018, "Social conflicts over extractivism in Latin America: concepts, theories and empirical evidence", *trAndeS teaching material*, no. 9, trAndeS - Postgraduate Program on Sustainable Development and Social Inequalities in the Andean Region, Berlin.

Schumpeter, JA 2010 [1942], *Capitalism, socialism and democracy*, with a new introduction by Joseph E Stiglitz, Routledge, London.

Slack, K 2009, 'The role of mining in the economics of developing countries: time for a new approach', in Richards, JP (ed), *Mining, society and a sustainable world*, Springer, Berlin.

Soria, L 2012, *Estudio sobre comunidades campesinas de Espinar y proceso de diálogo, negociación y consenso en actividades de exploración minera*, Final Report prepared for Oxfam, Societas Consultora de Análisis Social, Lima.

Soria, L 2014, *Evaluación de la inversión pública y privada en Espinar desarrollada en el marco del proyecto minero Tintaya*, Final Report prepared for Oxfam, Societas Consultora de Análisis Social, Lima.

Spalding, K 1974, *De indio a campesino: cambios en la estructura social del Perú colonial*, IEP, Lima.

Szablowski, D 2002, 'Mining, displacement and the World Bank: a case analysis of Compañía Minera Antamina's operations in Peru', *Journal of Business Ethics*, vol. 39, no. 3, pp. 247–73.

Szablowski, D 2007, *Transnational law and local struggles: mining communities and the World Bank*, Hart Publishing, Portland.

Taylor, L 1994, *Estructuras agrarias y cambios sociales en Cajamarca, siglos XIX–XX*, Editorial Martínez Compañón, Cajamarca.

Taylor, L 1997, 'La estrategia contrainsurgente, el PCP-SL y la Guerra Civil en el Perú, 1980–1996', *Debate Agrario*, vol. 26, pp. 81–110.

Thorp, R & Bertram, G 1978, *Peru 1890–1977: growth and policy in an open economy*, Columbia University Press, New York.

Torres, J (ed) 2013, *Los límites de la expansión minera en el Perú*, SER, Lima.

Tsing, A 2000, 'Inside the economy of appearances', *Public Culture*, vol. 12, no. 1, pp. 115–44.

Tsing, A 2005, *Friction: an ethnography of global connection*, Princeton University Press, Princeton.

Uccelli, F & García Llorens, M 2016, *Solo zapatillas de marca: jóvenes limeños y los límites de la inclusión desde el mercado*, IEP, Lima.

UNCTAD 2007, *Informe sobre inversiones en el mundo 2007: empresas transnacionales, industrias extractivas y desarrollo. Panorama general*, Naciones Unidas, New York/Geneva.

UNCTAD, WB & ICMM 2008, *Perú. Desafío de la riqueza mineral: utilizar la dotación de recursos para impulsar el desarrollo sostenible*, UNCTAD, Geneva.

United Nations 2010, *Achieving the millennium development goals with equality in Latin America and the Caribbean: progress and challenges*, United Nations Publications, Santiago de Chile.

Vega-Centeno, P 2006, 'De la barriada a la metropolización: Lima y la teoría urbana en la escena contemporánea', in DESCO, *Perú hoy: las ciudades en el Perú*, pp. 45–70, DESCO, Lima.

Vega-Centeno, P 2011, 'Los efectos urbanos de la minería en el Perú: del modelo de Cerro de Pasco y La Oroya al de Cajamarca', *Apuntes*, vol. 38, no. 68, pp. 109–36.

Velazco, J 1998, *Las actividades rurales no-agrarias en familias campesinas de la sierra norte del Perú*, Documento de trabajo no. 150, PUCP/CISEPA, Lima, viewed 23 October 2014, http://departamento.pucp.edu.pe/economia/images/documentos/DDD150.pdf

Viale, C & Monge, C 2012, 'La enfermedad chola', *Quehacer*, vol. 185, pp. 80–85.

Wachtel, N 1976, *Los vencidos: los indios del Perú frente a la conquista española (1530–1570)*, Alianza Editorial, Madrid.

Wallace, J 1984, 'Urban anthropology in Lima: an overview', *Latin American Research Review*, vol. 19, no. 3, pp. 57–85.

World Bank 2001, *Engendering development: through gender equality in rights, resources, and voice*, World Bank/Oxford University Press, New York, viewed 16 November 2012, www-wds.worldbank.org/external/default/WDSContentServer/WDSP/IB/2001/03/01/000094946_01020805393496/Rendered/PDF/multi_page.pdf

Zoomers, A 1999, *Linking livelihood strategies to development: experiences from the Bolivian Andes*, Royal Tropical Institute, Amsterdam.

Index

Note: **Bold** page numbers refer to tables.

Printed in the United States
by Baker & Taylor Publisher Services